TOPOLOGICAL METHODS IN THE THEORY OF FUNCTIONS OF A COMPLEX VARIABLE

By

MARSTON MORSE

Institute for Advanced Study
Princeton, New Jersey

PRINCETON
PRINCETON UNIVERSITY PRESS
LONDON: HUMPHREY MILFORD
OXFORD UNIVERSITY PRESS
1947

Copyright 1947
PRINCETON UNIVERSITY PRESS

Second printing 1951

Lithoprinted in U.S.A.
EDWARDS BROTHERS, INC.
ANN ARBOR, MICHIGAN
1951

FOREWORD

The following pages contain, in revised form, a set
of lectures given at Fine Hall in Princeton, New Jersey,
during the fall of 1945. A large part of the matter presented is the product of studies undertaken jointly by
the author and Dr. Maurice Heins, reference to which is
given in the bibliography. The first chapter on pseudo-
harmonic functions is, however, derived largely from the
author's paper on "The topology of pseudo-harmonic functions" Morse (1), while the fourth chapter on "The general order theorem" contains the first published proof of
the theorem there stated. The present exposition differs
from that in the joint papers, in that in the earlier
papers attention was focused on meromorphic functions and
the proofs then amended to include interior transformations. (See Stoilow (1), and Whyburn for previous work
on interior transformations. With Whyburn our transformations are interior and "light".) In these lectures
pseudo-harmonic functions and interior transformations
are the starting point, and the theorems specialize into
theorems on harmonic functions and meromorphic transformations.

The modern theory of meromorphic functions has distinguished itself by the fruitful use of the instruments
of modern analysis and in particular by its use of the
theories of integration. Its success along this line has
perhaps diverted its attention from some of the more
finitary and geometric aspects of function theory. His-
torically the geometric concepts of Riemann and Schwartz

contrast with the more arithmetical concepts of Weier-
strass and of the modern school.* The present lectures
seek to emphasize again the advantages of geometric
methods as a complement of other methods.

In the study of boundary values in a statistical
sense, significant finite topological properties of the
boundary images have been passed over, and the geometric
instruments appropriate for simple generalization not
always used. Passing to non-finitary aspects of the
theory, the critical points of a harmonic function on a
Jordan region, if infinite in number, stand in group
theoretic or topological relation to the boundary values,
assumed continuous, which arithmetic methods are not adequate to reveal. See Morse and Heins (1) III. On turn-
ing in still another direction of the theory, the topological development of pseudo-harmonic functions on the
basis of the topological characteristics of their contour
lines, makes the theory available, as Stefan Bergmann has
pointed out, for the study of problems in partial differential equations not otherwise reached.

However, it is not these negative aspects which are
most important but rather the possibility of attack on
new problems of a fundamental nature. One of these problems is the determination of properties of deformation
classes of meromorphic functions with prescribed zeros,
poles and branch points. See Morse and Heins (2). Here
a connection is made between the interest of the topolog-
ist in homotopy theories, and the classical interest in
theorems on normal families, or covering theorems of the
Picard type.

These lectures form merely the beginning of studies
of this type. It is hoped that they may strike a respon-
sive chord in the hearts of those to whom there is an
appeal in the geometric approach.

---

*The remarkable work of Lars Ahlfors should be excepted.

TABLE OF CONTENTS

TABLE OF CONTENTS

CHAPTER I

PSEUDO-HARMONIC FUNCTIONS

§1. Introduction

We shall consider meromorphic functions $F(z)$ on a region G (open) bounded by v Jordan curves

$$(1.1) \qquad (B_1, \ldots, B_v) = (B).$$

We shall suppose that $F(z)$ is defined on $\overline{G}$ (the closure of G), and is analytic on G except for poles, and continuous at points of (B). The number of poles of $F(z)$ on G is necessarily finite.

Alongside of $F(z)$ we shall consider <u>interior transformations</u> $w = f(z)$ of G into the w-sphere. Such transformations are generalizations of meromorphic functions. To define such a transformation one begins with a definition of an interior transformation in the neighborhood of an arbitrary point z_0 of G. Suppose that $F(t)$ is a non-constant, analytic function defined on a neighborhood N of t_0. One subjects N to a 1 - 1 continuous sense-preserving transformation

$$(1.2) \qquad t = \phi(z) \qquad\qquad (t_0 = \phi(z_0))$$

which maps N onto a neighborhood N_1 of z_0. The function

$$(1.3) \qquad F(\phi(z)) = f(z)$$

thereby defined on N_1 is called an interior transformation $w = f(z)$ of N_1 into the w-sphere. A transformation
$w = f(z)$ defined on G will be termed an <u>interior</u> <u>trans-
formation</u> of G if $w = f(z)$ is an interior transformation
of some neighborhood of each point of G.

　　We shall admit the possibility that F(t) have a
pole at t_0 and then say that $f(z)$ in (1.3) is an interior
transformation <u>with</u> <u>a</u> <u>pole</u> at z_0. We shall consider interior transformations with at most a finite number of
poles on G, and suppose that $f(z)$ is defined on $\overline{G}$ and continuous at points of (B). We do not say that $f(z)$ is an
interior transformation on the boundary (B), although it
is clear that $f(z)$ might in certain cases be extended in
definition so as to be an interior transformation of a
neighborhood of each boundary point.

　　We add an example of an interior transformation.
Let F(t) be an arbitrary polynomial in t. Set $z = $
$x + iy$. Replace t in F(t) by

$$t = 2x + iy = \phi(z).$$

The resulting function $F(\phi(z)) = f(z)$ will be interior
but not analytic.

　　Interior transformations have been introduced at the
very beginning not because they are our principal object
of study but because they furnish a convenient medium for
illustrating the new topological methods. The zeros,
poles, and branch points of $f(z)$ are a fundamental source
of study in the classical theory of functions. What are
the relations between their numbers under given boundary
conditions? To what extent do they determine the meromorphic function either with or without a knowledge of the
boundary values? Theorems of this type have been given by
Radó, Stoïlow, Walsh, Backlund, Lucas and others. Possibly the simplest of these theorems is that of Lucas,
as follows. If P(z) is any polynomial in z, the zeros of

P'(z) are found in any convex region which contains the
zeros of P(z). Many of the theorems of the above authors
have their generalizations for interior transformations.

   We have referred to branch points. It is necessary
to give this term a meaning in the case of interior transformations. As is well known, a non-constant meromorphic
function f(z) if restricted to a sufficiently small
neighborhood of a point z_0, takes on every value w in a
sufficiently small neighborhood of $w_0 = f(z_0)$ an integral
number of times m, w_0 alone excepted. If $m > 1$, the inverse of f(z) is said to have a <u>branch point</u> of order
m - 1 at the point w_0. With the neighborhood of z_0 restricted as above, f(z) defines a <u>meromorphic element</u>.
Any interior transformation obtained from a meromorphic
element by a homeomorphic change of independent variable
will be called an <u>interior element</u>. The totality of
function values w remains unaltered. The neighborhood of
w_0 is covered the same number m of times by the interior
element as by the defining meromorphic element. It is
therefore appropriate to say that the interior element
defines a <u>branch point</u> of order m - 1 at w_0 whenever
$m > 1$. It is clear that this branch point order depends
only on the given interior element and does not vary
with the various meromorphic elements which may be used
to define it. The orders of zeros or poles of an interior element are similarly defined as the orders of the
zeros or poles of defining meromorphic elements.

   <u>Methods</u>. The definition of an interior transformation is such that f'(z) does not exist in general. The
classical use of the Cauchy integral

$$\frac{1}{2\pi i} \int_C \frac{f'(z)}{f(z)} \, dz$$

to find the difference between the number of zeros and
poles of f(z) within C is thus unavailable, at least in

any a priori sense. Branch points at ordinary points
cannot be located in general as zeros of $f'(z)$. In the
classical theory $f'(z)$ is either null or infinite within
C, or defines a direction represented by arc $f'(z)$.
Vector methods can then be used to locate the zeros of
$f'(z)$, as in the case of one of the proofs of the fundamental theorem on algebra. These vector methods fail
in the general theory, at least in the absence of some
effective change of independent variable in the large.
More important are positive advantages of topological
methods. The classical treatment of boundary values by
means of an integral in general ignores extremal properties of boundary values, such for example as the extremal
values of $|f(z)|$. The images g_i under $w = f(z)$ of the
boundary curves B_i, if locally simple, have important
topological properties which more than compensate for the
lack of derivatives. (A closed curve g is termed locally simple if it is the continuous and locally 1 - 1
image of a unit circle.)

In a final section we shall introduce a deformation
theory of interior or meromorphic functions, considering
one-parameter families of such functions

$$w = F(z, t) \qquad\qquad (0 \leq t \leq 1)$$

where for each t, $F(z, t)$ is an interior transformation
defined on G, and such that the point w varies continuously on the "extended" w-plane with both z and t. Such
a one-parameter family of interior transformations will
be termed a deformation of $F(z, 0)$ into $F(z, 1)$. We admit deformations in which the zeros, poles and branch
point antecedents are held fast, and put functions $f(z)$
which can be thus deformed into each other, into the same
restricted deformation class. Deformations are also admitted in which the number but not the position of the
zeros, poles, and branch point antecedents are held fast.

Topological invariants of the admissible deformations
have been determined which characterize the deformation
classes whether restricted or unrestricted. A question
of great interest is whether the deformation classes
defined by a use of meromorphic functions alone are
identical with those defined when the more general interior transformations are used. Details will not be given.
For proofs see Morse and Heins (2).

In general one seeks to distinguish those basic
theorems on meromorphic functions which can be established for meromorphic functions but not for interior
transformations. One such theorem is the Liouville theorem that a function which is analytic in the finite
z-plane and bounded in absolute value, is constant. This
is not true if stated for interior transformations. One
can indeed map the finite z-plane homeomorphically on the
interior of the circle $|w| < 1$ by the interior transform-
ation $w = z/(1 + |z|)$; defined for every finite z.
Clearly $f(z)$ is not constant. On the other hand, we shall
see that many theorems hold equally well for meromorphic
functions and interior transformations.

§2.   Pseudo-harmonic functions

The study of meromorphic functions leads naturally
to harmonic functions. In a similar manner the study of
interior transformations leads to functions which we shall
call pseudo-harmonic and shall presently define.

We begin by considering the function

$$(2.0) \qquad\qquad U(x, y) = \log |f(z)|$$

in case $f(z)$ is meromorphic. As is well known this is
the real part of $\log f(z)$ and is accordingly harmonic
whenever the continuous branches of $\log f(z)$ are analytic.
Thus $U(x, y)$ is harmonic at every point $z = x + iy$ not a
zero or pole of $f(z)$. Let $z = a$ be a zero or pole of

f(z). Then f(z) admits a representation

$$f(z) = (z - a)^m A(z) \qquad (A(a) \neq 0)$$

where A(z) is analytic at z = a. Neighboring z - a, U thus has the form

$$m \log |z - a| + \omega(x, y)$$

where $\omega(x, y)$ is harmonic. The function U has a <u>logarithmic pole</u> at z = a. More generally one considers harmonic functions of the form

$$k \log |z - a| + \omega(x, y) \qquad (k \neq 0)$$

where k is real but not necessarily an integer.

The <u>critical points</u> of U in (2.0) in the ordinary sense are the points at which $U_x = U_y = 0$. By virtue of the Cauchy-Riemann differential equations, when $f(z) \neq 0$ each such critical point is a zero of

$$\frac{d}{dz} \log f(z) = \frac{f'(z)}{f(z)} ,$$

and is thus a zero of f'(z). Thus the zeros and poles of f(z) are reflected by the logarithmic poles of U(x,y) and the zeros of f'(z) by the critical points of U.

Before coming to the definition of a pseudo-harmonic function, it will be helpful to give a description of the level arcs through a given point (x_0, y_0) of a non-constant harmonic function U. We are concerned with the locus

(2.1) $$U(x, y) - U(x_0, y_0) = 0$$

neighboring (x_0, y_0). The harmonic function U is the real part of an analytic function $f(z)$. If $z_0 = x_0 + iy_0$, $f(z) - f(z_0)$ vanishes at z_0 and has the form

$$(2.2) \qquad f(z) - f(z_0) = (z - z_0)^m A(z) \qquad (A(z_0) \neq 0.)$$

We shall make a conformal transformation of a neighborhood of z_0 following which the desired level curves will appear as straight lines. This conformal transformation has the form

$$(2.3) \qquad w = (z - z_0)A^{1/m}(z),$$

where any continuous single-valued branch of the mth root may be used. The transformation (2.3) is locally 1 - 1 and conformal neighboring z_0, since at z_0

$$\frac{dw}{dz} = A^{1/m}(z_0) \neq 0.$$

In terms of the variable w,

$$f(z) - f(z_0) = w^m.$$

If $w = u + iv$ the required level lines are the level lines through the origin of

$$R(u + iv)^m \qquad (R = \text{Real part})$$

for example, if $m = 2$, the level lines of $u^2 - v^2$. If (r, θ) are polar coordinates in the w-plane

$$w^m = r^m(\cos m\theta + i \sin m\theta).$$

Thus by virtue of the transformation from (x, y) to (u, v) to (r, θ)

$$(2.4) \qquad U(x, y) - U(x_0, y_0) = r^m \cos m\theta.$$

In the (u, v) plane the required level lines are rays on which $\cos m\theta = 0$. There are 2m of these rays, each making an angle of $\frac{\pi}{m}$ with its successor. For example, if m = 1, the directions are $\frac{\pi}{2}, \frac{3\pi}{2}$. If m = 2 the directions are

$$\frac{\pi}{4}, \frac{3\pi}{4}, \frac{5\pi}{4}, \frac{7\pi}{4},$$

that is, the lines of slope ±1. Since our transformation from the (x, y) plane to the (u, v)-plane was conformal, it follows that the level curves through (x_0, y_0) consist of m curves without singularity, each making an angle of $\frac{\pi}{m}$ at (x_0, y_0) with its successor. Another way of putting this result follows.

THEOREM[*] 2.1. Let (x_0, y_0) be a point at which U is harmonic. Suppose U is not constant. There exists an arbitrarily small neighborhood N of (x_0, y_0) whose closure is the homeomorph of a plane circular disc such that (x_0, y_0) corresponds to the center of the disc and the locus

$$(2.5) \qquad U(x, y) - U(x_0, y_0) = 0$$

corresponds to a set of 2m rays leading from the disc center and making successive sectors of central angle $\frac{\pi}{m}$. As a variable point crosses any one of these level lines (except at (x_0, y_0)) the difference (2.5) changes sign.

---

*Theorem 2.1 stated for pseudo-harmonic functions will be termed Theorem 2.1a.

The first statement of the theorem is an immediate consequence of the mapping of the (x, y)-plane into the (u, v)-plane as above. One chooses the disc $r \leqq r_0$ in the (u, v)-plane with r_0 so small that the mapping of the (x, y)-plane into the (u, v)-plane is 1 - 1 and conformal for $r \leqq r_0$. The second statement of the theorem follows from (2.4) and the fact that cos mθ changes sign with increasing θ whenever it vanishes.

With U non-constant, the smallest value of m in the theorem is 1, in which case there is but one non-singular level curve through (x_0, y_0). A particular consequence of the theorem is that U can never assume a relative maximum or minimum at a point (x_0, y_0) neighboring which it is harmonic. For one sees that $U(x, y) - U(x_0, y_0)$ is both positive and negative in every neighborhood of (x_0, y_0).

Definition of pseudo-harmonic functions. Let u(x, y) be a function which is harmonic and not identically constant in a neighborhood N of a point (x_0, y_0). Let the points of N be subjected to an arbitrary sense-preserving homeomorphism T in which N corresponds to another neighborhood N' of (x_0, y_0) and the point (x, y) on N corresponds to a point (x', y') on N'. It will be convenient to suppose that (x_0, y_0) corresponds to itself under T. Under T set

(2.6)   u(x, y) = U(x', y').

The function U(x', y') will be termed pseudo-harmonic on N'. This definition will be extended to the case where u(x, y) has a logarithmic pole at (x_0, y_0). In this case

$$u(x, y) = k \log |z - z_0| + \omega(x, y) \quad (k \neq 0)$$

where ω(x, y) is harmonic in a neighborhood of (x_0, y_0).

Under the above homeomorphism T, relation (2.6) defines
what is termed a pseudo-harmonic function with logarithmic pole at (x_0, y_0). More generally, we shall admit functions $U(x, y)$ which are pseudo-harmonic, except for logarithmic poles, in some neighborhood of every point of
the region G and are continuous on the boundary of G.

With the above definition of a pseudo-harmonic function, it is clear that the level curves of a function U
which is pseudo-harmonic in the neighborhood of a point
(x_0, y_0) of G are such that Theorem 2.1a holds (i.e.,
Theorem 2.1 with "pseudo-harmonic" replacing "harmonic").
As a corollary it follows that a pseudo-harmonic function
assumes a finite relative maximum or minimum at no point
of G.

§3. Critical points of U on G.

Points of G at which $U < c$ will be said to be below
c; those at which $U > c$, above c. Let (x_0, y_0) be a
point of G not a logarithmic pole and set

$$U(x_0, y_0) = c.$$

Refer to Theorem 2.1a. This theorem gives a canonical
representation of the level arcs of U ending at (x_0, y_0).
The neighborhood N of (x_0, y_0) of Theorem 2.1a will be
termed canonical. Any one of the open, connected subsets
of N bounded by two successive arcs at the level c and
the intercepted arc of the boundary of N will be called a
sector of N. There are m sectors of N below c, and m
sectors above c. If m = 1 the point (x_0, y_0) will be
termed ordinary, otherwise critical. When $m > 1$ the number m - 1 will be called the multiplicity of the critical
point (x_0, y_0) of G. For our purposes the essential topological characteristic of these critical points is the
existence of two or more sectors of a canonical neighbor-

hood below $U(x_0, y_0)$. We shall find that there are
boundary points with this same characteristic. Such critical points will be called _saddle_ _points_ to distinguish
them from other types of critical points such as points
of relative minimum of U on (B).

When $U(x, y)$ is harmonic, a necessary and sufficient
condition that a point (x_0, y_0) of G be critical is that,
at (x_0, y_0),

$$U_x = U_y = 0.$$

In fact we have seen that if U is the real part of $f(z)$,
and $f(z) - f(z_0)$ has a zero of the mth order, there are
just 2m level rays of $U(x, y)$ tending to (x_0, y_0). Thus
the order $m - 1$ of vanishing of $f'(z)$ is the multiplicity
$m - 1$ of (x_0, y_0) as a critical point of U.

At a _boundary_ point of G the partial derivatives
U_x, U_y of a function harmonic on G do not in general
exist. This is true of pseudo-harmonic functions at all
points of G. Thus classical methods are inadequate to
characterize the boundary points of G as critical points
of $f(z)$ both in the case where U is harmonic and pseudo-
harmonic on G.

§4. Critical points of U on (B)

Examples. Since we are dealing with pseudo-harmonic
functions no generality will be lost if we subject $\overline{G}$ to
a homeomorphism T by virtue of which the boundary Jordan
curves become circles. This is on the supposition that
the values of U are taken equal at points which correspond under T. We suppose accordingly that each boundary
curve is a circle. Let B_1 be a boundary circle and let
$U^{(1)}$ be the function defined by $U(x, y)$ on B_1.

Boundary conditions[*] A. We shall assume that $U^{(i)}$
has at most a finite number of points of relative extremum on B_i (i = 1, 2, ... v).

Between these points of extremum $U^{(i)}$ is monotonic-
ally increasing or decreasing on B_i. As a consequence of
this we shall show in the next section that a sensed
level arc of U is either closed or leads on continuation
to a point on (B). We shall show in §5 that the number of
level arcs which terminate at any one boundary point is
finite, and that the number of boundary points at which
more than one level arc terminates is finite. We shall
continue in the present section with a number of examples
illustrating the various possibilities.

Example 1. Let G be a region on which y > 0, bounded
by a circle tangent to the x-axis at the origin. Let U
be the real part of z^3, that is the function

$$U(x,\ y) \equiv x(x^2 - 3y^2).$$

The level lines of U through the origin are the y axis
and the lines x = $\pm \sqrt{3y}$. Thus the negative x-axis lies
in a sector on which U < 0 and the positive x-axis in a
sector on which U > 0. It is readily seen that U increases on the boundary circle B as the origin is passed
with increasing x. The origin is thus not an extremum
of the boundary function defined by U. On a neighborhood
of z = 0 relative to G consisting of points on G within
a sufficiently small distance of z = 0, the subsets of
points of G below c lie on two distinct components (connected sets) or sectors. The origin will then be called
a saddle point of multiplicity 1. This example shows
that a boundary point which affords no extremum to $U^{(i)}$
can be a saddle point in the above sense.

---

[*]Various less restrictive boundary conditions will be
treated later.

Example 2. Boundary points at which U assumes a
relative minimum will be regarded as critical points. To
illustrate such a point, let G be the region of Example 1
and let $U \equiv y$ on G. U is always positive, and on $\overline{G}$ has a
minimum 0 at the origin. We regard the origin as a crit-
ical point of U.

If, however, $U \equiv -y$, U assumes a relative maximum at
the origin and we do not regard the origin as a critical
point of U. The reason for this distinction between maximum and minimum points on (B) is as follows. Let U_c
represent the point set on which $U \leq c$. As c increases
through a minimum value of U, the number of connected
pieces of U_c increases by 1. That is, U_c changes its
topological character. However, as c increases through a
maximum value of U, there is on that account no topological change in U_c. Neighboring a boundary point (x_0, y_0)
of relative maximum, U_c enlarges as c increases to
$U(x_0, y_0)$, and when $c \geq U(x_0, y_0)$ there is no change in
U_c neighboring (x_0, y_0) as c increases. A strict analysis of this situation will appear in a later section.

Example 3. A boundary point may be a saddlepoint of
-U but not of U. Let $U \equiv x^2 - y^2$ on the region G of
Example 1. On the circular boundary B_1 of G it appears
that $U^{(1)}$ has a relative minimum at $z = 0$, but that
$z = 0$ is not a point of relative minimum of U on $\overline{G}$. There
are two level curves tending to $z = 0$ on G. On any circular neighborhood N of $z = 0$, relative to G, of sufficiently small radius the subset of points below 0 consists
of just one component. By virtue of the general definition of boundary critical points as we shall give it,
$z = 0$ is then an ordinary point. Note that U has a critical point at the origin, in the classical sense. On
this same region, -U will give rise to two sectors below
0 on arbitrarily small circular neighborhoods of $z = 0$.
The origin will accordingly be regarded as a saddle point
of -U but not of U.

These examples show that the topological concept of
critical point is "_relative_" both to U and to G.

In our terminology the critical points of U shall
include the interior and boundary saddle points of U, together with the points of relative minimum of U. Until
§12 is reached, we shall assume that there are no logarithmic poles. When the case in which there are no logarithmic poles has been fully treated there is a simple
device which reduces the case of a logarithmic pole to
the case already treated.

To distinguish between critical points as used in
the above sense, and critical points as defined by the
vanishing of the first partial derivatives, one might
call the latter differential critical points and the former topological critical points. In these lectures we
shall drop the adjective "topological".

§5. Level arcs leading to a boundary point z_0 not an extremum point of U.

This section is concerned with the level curves of
U in the neighborhood of z_0. To that end it is desirable
to introduce a simple neighborhood D of z_0. Suppose
that z_0 is on the boundary circle B_1. Set

$$(5.1) \qquad\qquad U(x_0,\, y_0) = c. \qquad (z_0 = x_0 + iy_0)$$

Let D be the intersection with G of an open circular disc
centered at z_0 and such that $\overline{D}$ intersects no boundary
circle (B) other than B_1, and that on the intersection ω
of B_1 and $\overline{D}$, U = c only at z_0.

The final condition on D can be fulfilled by virtue
of the assumption that $U^{(1)}$ has at most a finite number
of points of extremum on B_1.

Each point of D at the level c has a canonical
neighborhood N in the sense of Theorem 2.1a. By a cross

arc of such a neighborhood will be meant an arc of N at
the level c, given by the sum of diametrically opposite
rays in Theorem 2.1a. If a point z of D is ordinary,
there is just one such cross arc on N. If the point z is
a saddle point of multiplicity μ, there are $\mu + 1$ such
cross arcs.

Note that an arc at the level c on D is necessarily
simple. Otherwise it would bound one or more regions R
on D. On any such region R, U would not be constant by
definition of a pseudo-harmonic function, and would accordingly assume an extremum value at some point of R.
This is impossible as we have seen.

We are concerned with the "continuation" of an arc
at the level c.

Maximal arcs at the level c. The locus U = c on any
closed subset of D can be covered by a finite set of
canonical neighborhoods of the type of Theorem 2.1a.
Accordingly, the locus U = c on D as a whole can be
covered by a countable set of canonical neighborhoods

$$N_n \qquad\qquad (n = 1, 2, \ldots)$$

of points P_n on D with the following properties. The
points P_n are distinct and the sets $\overline{N}_n$ are on D. Any
given closed subset of D intersects at most a finite subset of the neighborhoods N_n. The diameter of the neighborhood N_n tends to 0 as P_n tends to the boundary of D.

An open simple arc g of D at the level c will be
termed maximal if it is the sum of cross arcs of a subset of neighborhoods N_n and has no first or last cross
arc. The existence of such a maximal arc g containing a
preassigned cross arc follows from the fact that the end
points of any given cross arc k are on another cross arc
(of the given set) which intersects k in a simple arc.
This process of enlarging a simple arc which is the sum
of a finite number of cross arcs will lead by a countable
number of steps to an open simple arc g which is maximal

in the above sense. An open simple arc such as g can be
represented in the form

(5.2) $z = a(t)$ $(0 < t < 1)$

as the 1 - 1 continuous image of the t interval. The
basic properties of these maximal arcs g at the level c
are as follows.

Let ω be the boundary of D on the circle B_1, and ω_1
the remaining boundary of D.

(a) As t tends to 0 or 1, the distance of a(t) from
$\omega + \omega_1$ tends to 0.

This is an immediate consequence of the fact that g
is simple and accordingly never enters a neighborhood N_m
twice, while in any infinite sequence of different neighborhoods N_m, the distance of N_m from $\omega + \omega_1$ tends to 0
as m becomes infinite.

(b) When t tends to 1, a(t) tends to ω, or else ω_1
(not both), and if a(t) tends to ω it tends to z_0. Similarly when t tends to 0.

Any limit point on ω of points of g must lie at
the level c and hence coincide with z_0. Points on ω_1 at
the level c cannot be connected to z_0 at the level c,
among points arbitrarily close to $\omega + \omega_1$. Otherwise
there would be a whole boundary arc of ω at the level c.
If a(t) tends to ω as t tends to 1, it cannot also tend
to ω_1. It accordingly tends to z_0.

(c) If a(t) tends to z_0 as t tends to 1, then a(t)
tends to ω_1 as t tends to 0. Similarly on interchanging
1 and 0.

Otherwise g could be closed by adding its two limit-
ing end points at z_0 and hence bound a region on D, which
is impossible.

For similar reasons it is clear that two maximal
arcs g both of which have an end point at z_0 cannot intersect in any other point of D.

(d) <u>There</u> <u>are</u> <u>at</u> <u>most</u> <u>a</u> <u>finite</u> <u>number</u> <u>of</u> <u>maximal</u> <u>arcs</u> g <u>with</u> <u>an</u> <u>end</u> <u>point</u> <u>at</u> z_0.

Let K be the intersection with D of a circle with center at z_0 and with a radius less than the maximum radius of D drawn from z_0. If a(t) tends to z_0 as t tends to 1 (or 0) it follows from (c) that a(t) tends to ω_1 as t tends to 0 (or 1), so that g intersects K. The points on K at the level c form a closed set S on D since the intersection of K with (B) is not at the level c. The points of S can be covered by a finite number of canonical neighborhoods N_n on D possessing a finite number of cross arcs. Since maximal arcs on D are simple, the number of such arcs intersecting K is at most the number of such cross arcs and accordingly finite.

(e) <u>There</u> <u>is,</u> <u>at</u> <u>least</u> <u>one</u> <u>maximal</u> <u>arc</u> g <u>on</u> D <u>with</u> <u>an</u> <u>end</u> <u>point</u> <u>at</u> z_0.

The point z_0 is not an isolated point at the level c since z_0 is not an extremum point of U. There is accordingly a sequence z_n of points at the level c on D which tend to z_0 as n becomes finite. If infinitely many of these points lie on any one maximal arc g, g must have an end point at z_0. This is the only possibility. Otherwise there would be infinitely many maximal arcs g_n with points arbitrarily near z_0. These arcs would tend to ω_1 as t tends to 0 or 1, and would intersect K. But we have seen that there are at most a finite number of maximal arcs intersecting K; (e) follows.

(f) <u>Points</u> <u>at</u> <u>the</u> <u>level</u> c <u>not</u> <u>on</u> <u>maximal</u> <u>arcs</u> <u>with</u> <u>an</u> <u>end</u> <u>point</u> <u>at</u> z_0 <u>are</u> <u>bounded</u> <u>away</u> <u>from</u> z_0.

The proof of (e) implies this fact.

§6. <u>Canonical</u> <u>neighborhoods</u> <u>of</u> <u>a</u> <u>boundary</u> <u>point</u> z_0
<u>not</u> <u>an</u> <u>extremum</u> <u>point</u> <u>of</u> U

Let the <u>positive</u> extrinsic sense of a plane Jordan curve g be that sense in which the order of interior points is 1 with respect to g.

LEMMA 6.1. Let T be an arbitrary homeomorphism of a Jordan curve g_1 onto a second Jordan curve g_2 which preserves the extrinsic sense. There exists a homeomorphism of the closed domain S_1 bounded by g_1 onto the closed domain S_2 bounded by g_2 which agrees with T on g_1.

That some sense preserving homeomorphism of S_1 onto S_2 exists, is well known. To verify that this homeomorphism can be prescribed on the boundary subject to the condition of preservation of the extrinsic sense of the boundary, one merely has to note that the lemma is true when S_1 and S_2 reduce to the same circular disc H. If r and θ are polar coordinates with pole at the center of H, T can be represented by a continuous and increasing function F(θ), with F(θ) ≡ F(θ + 2π). The transformation

(6.1)   θ' = F(θ), r' = r,

defines the required homeomorphism of S_1 onto S_2 in the case $S_1 = S_2 = H$. In the general case one first maps S_1 and S_2 onto a circular disc, then maps this disc onto itself with the prescribed mapping on the boundary. The required mapping of S_1 onto S_2 is obtained from these mappings.

To return to the principal problem refer to the neighborhood D of z_0 as described in the preceding section. (Cf. §5) Let

(6.2) $h_1, \ldots, h_n$

be the "maximal arcs" on D which approach z_0. Recall that these arcs do not intersect on D other than in their end point z_0. Let

(6.3) $P_1, \ldots, P_n$

be their end points on the boundary ω_1 of D not on (B).
Recall that ω_1 is a circular arc at a constant distance
from z_0. The fact that the arc h_j approaches a definite
P_j on ω_1 follows from the fact that the points on ω_1 at
the level c of z_0 can be covered by a finite number of
canonical neighborhoods N_m with a corresponding finite
number of cross arcs.

Let P_0 and P_{n+1} be the end points of ω_1 on the
boundary B_1 of G, and suppose that the notation is such
that the points (see Fig. 1)

(6.4) $P_0, P_1, \ldots, P_n, P_{n+1}$

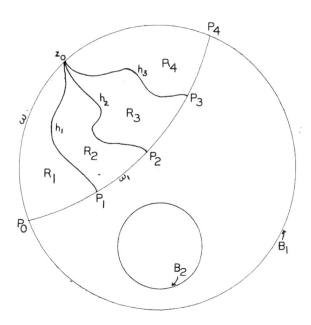

Figure 1.   (n = 3)

appear on the boundary of D in the order defined by its positive sense.

It follows from the preceding lemma that there exists a <u>homeomorphism</u> T between $\bar{D}$ and a semi-disc H

$$(6.5) \qquad\qquad (r \leqq 1, \; 0 \leqq \theta \leqq \pi)$$

in which z_0 corresponds to the center $r = 0$ of H, the arc of $\bar{D}$ on B_1 corresponds to the diameter of H, while the arc h_j corresponds to a ray on H on which

$$(6.6) \qquad\qquad \theta = \frac{j\pi}{n+1} \qquad\qquad (j = 1, \; \ldots, \; n).$$

The arcs h_j divide D into a sequence of disjoint regions,

$$R_1, \; R_2, \; \ldots, \; R_{n+1},$$

with z_0, P_{j-1} and P_j on the boundary of R_j. One first maps each arc h_j homeomorphically onto the corresponding ray (6.6). Consistently with this (Lemma 6.1) one then maps $\bar{R}_j$ homeomorphically onto the jth of the domains into which the semi-disc H is divided by the rays (6.6). This gives the desired homeomorphism T.

In the representation of $\bar{D}$ by H the points at the level c other than those on the rays (6.6) are bounded from $r = 0$, so that a suitable semi-disc $r \leqq r_1 < 1$ of H will represent no points of $\bar{D}$ at the level c other than on arcs tending to z_0. Hence one has the principal theorem.

THEOREM 6.1. <u>Let</u> z_0 <u>be a point on</u> (B) <u>which is not isolated among points at its level</u> c. <u>There exists a canonical neighborhood</u> N <u>of</u> z_0 <u>relative to</u> $\bar{G}$ <u>whose closure is the homeomorph of a semi-disc</u> H <u>such that</u> z_0 corres-

ponds to the center 0 of H, the intersection of $\bar{N}$ with
(B) corresponds to the diameter of H, while the points of
N at the level c are represented by n rays (n > 0) emanating from 0. These rays divide N into n + 1 regions or
sectors on which U - c alternates in sign. Each point
other than z_0 on these rays represents an ordinary point
on the locus U = c.

A boundary point z_0 which is not isolated among
points at its level c will be called ordinary if there
is at most one sector below c in a canonical neighborhood
N of z_0. Otherwise z_0 will be called a critical point or
saddle point of multiplicity m - 1, where m is the number
of sectors of N below c.

An interior point P at the level c of G will be
called a multiple point if there is more than one arc of
the locus U = c through P. An interior point is a multiple point if and only if it is a saddle point.

It is otherwise for boundary points. A boundary
point P will be called a multiple point if there is more
than one arc of the locus U = c tending to P. As has
been seen in examples, there can be multiple points on
(B) which are not saddle points of U. Naturally every
saddle point on (B) is a multiple point. One sees at
once that every multiple point P on (B) is a saddle point of
U or of -U, and of both U and -U if the number of sectors
in a canonical neighborhood of P exceeds three.

§7. Multiple points.

The basic theorem here is the following:

THEOREM 7.1. The set of multiple points of the
level lines is isolated.

An interior multiple point P is isolated, for if coordinates in terms of which U is harmonic are chosen in

a neighborhood of P then the multiple points are differential critical points of U. The subset of multiple
points at any <u>one</u> level c is also isolated, due to the
nature of the locus U = c in a canonical neighborhood of
a boundary point. Thus a boundary point z_0 at the level
c could be a limit point of multiple points only if it
were a limit point of multiple points not at the level c.
We may confine ourselves to the case where z_0 is a limit
point of multiple points below the level c.

To continue, one needs the notion of a U-trajectory.
By a U-<u>trajectory</u> we shall mean any simple arc on which U
is increasing or decreasing. The existence of a U-trajectory through any ordinary interior point P of G is
immediately demonstrated. One selects, in a neighborhood
of P, coordinates (x, y) in terms of which U is harmonic,
arranging it so that $U_x \neq 0$ at P = (x_0, y_0). Clearly the
arc through (x_0, y_0) on which y = y_0 is a U-trajectory.
If P is a boundary point, any sufficiently short arc of
the boundary (B) terminating at P is a U-trajectory.

LEMMA 7.1. <u>The multiple points of the level lines</u>
<u>are bounded away from any boundary point</u> z_0 <u>which is not</u>
<u>an extremum point of</u> U.

To prove this, let R be a sector of a canonical
neighborhood of z_0. Such sectors are of two types: an
ordinary type in which there are two arcs on the boundary
of R at the level c of z_0 and which terminate at z_0, and
a <u>boundary</u> type in which there is just <u>one</u> such arc on
the boundary of R.

Case I. <u>The sector R is not of boundary type</u>.
Without loss of generality we can suppose that R is
below the level c of z_0. We seek to prove that there are
no multiple points on R below c clustering at z_0. Let h_1
and h_2 be the two boundary arcs of R at the level c terminating at z_0. Let inner points P_1 and P_2 of h_1 and h_2
respectively, be joined on R by a simple arc k which is a

U-trajectory neighboring P_1 and P_2 respectively. The arc k will divide R into two regions of which one, R_1, will contain z_0. If e is a sufficiently small positive constant, there will be just two points Q_1 and Q_2 on the boundary of R_1 at the level c-e lying respectively on sub-arcs of k neighboring P_1 and P_2. Hence there can be but one simple "maximal arc" on R_1 at the level c-e. For the end points of such an arc must lie at Q_1 and Q_2 respectively, and the existence of a second such arc w implies the existence of a region on R_1 bounded by arcs at the level c-e. There can accordingly be no multiple points on R_1 at the level c-e, for the existence of such a multiple point implies the existence of two or more simple maximal arcs at the level c-e.

Thus multiple points do not cluster at z_0 on R in Case I.

Case II. The sector R is of boundary type.

As in Case I one supposes that R is below c. The arc h_1 is chosen as in Case I as a boundary arc of R at the level c terminating at z_0, and P_1 is taken as an inner point of h_1. Let P_1 be joined on R by a simple arc k to any boundary point of R below c, with some sub-arc of k neighboring P_1 a U-trajectory. The arc k again divides R into two regions of which, R_1, will contain z_0. If e is a sufficiently small positive constant there will again be just two points Q_1 and Q_2 on the boundary of R_1 at the level c-e, and the proof is concluded as in Case I.

This establishes Lemma 7.1.

To establish Theorem 7.1 in general we must still consider neighborhoods of boundary points z_0 which are points of relative extrema of U. It will be sufficient to consider the case in which z_0 affords a relative maximum to U.

Any sufficiently small neighborhood N of z_0 relative to G will be below c, with $\overline{N}$ below c except at z_0. If e is a sufficiently small positive constant the set of

points on the boundary of N at the level $c-e$ will consist
of just two points on (B). It follows as in the proof of
Lemma 7.1 that the multiple points of the level lines do
not cluster at z_0. We thus have a result complementing
the preceding lemma:

LEMMA 7.2. <u>The multiple points of the level lines
of U are bounded away from any boundary point which is a
point of relative extremum of U.</u>

Theorem 7.1 now follows from Lemmas 7.1 and 7.2.

The proofs of the preceding two lemmas enable us to
infer another result of use later.

THEOREM 7.2. <u>Let z_0 on (B) be a point of relative
extremum of U. There exists a canonical neighborhood N
of z_0 whose closure is homeomorphic with a semi-disc,
such that z_0 corresponds to the center of the disc, the
boundary of N which is on (B) corresponds to the straight
edge of the semi-disc; the boundary of N which is on G is
at a level $c_1 \neq c$ such that the points of N, excepting z_0,
all lie between the levels c and c_1.</u>

From the isolated character of the multiple points
of the level lines of U one infers that the number of
such multiple points is finite. Hence the number of
saddle points of U (or of -U) is finite. Any such saddle
point is called a <u>critical point</u> of U. Points of relative minimum of U, but not points of relative maximum,
are also termed critical points of U. Thus the critical
points of U are finite in number. The function -U has
the same interior critical points as U, but does not
agree with U in all its critical points on (B).

We shall suppose that each canonical neighborhood N
of a point z_0 on G is so restricted in diameter that,
with z_0 alone excepted, there are no critical points of U
or -U on $\overline{N}$.

§8. The sets U$_c$ and their maximal boundary
       arcs w at the level c.

The set of points at which U is at most c will be
denoted by U$_c$. Such a set is closed. Its boundary will
include the set of points, denoted by [U = c], at which
U = c. It will also include any arcs of the boundary (B)
of G which are below c.

A maximal boundary arc w of U$_c$ at the level c will
now be defined. Any simple boundary arc of U$_c$ at the
level c can be unambiguously continued until it enters a
canonical neighborhood N of a boundary point[*] P or of a
saddle point[*] P of G. In either case the arc enters N at
the level c on the boundary h of a sector S of N below c.
If P is a boundary point and if S is a sector of boundary
type (cf. §6), the arc w shall end at P. Otherwise the
arc shall continue with the other boundary arc of S at the
level c. So continued there will result an arc W which
is either a cycle or else has its two end-points on (B).

Such an arc may have multiple points at saddle points
of U. At such multiple points two or more locally simple
branches of w have the multiple point in common without
crossing. The type of continuation which defines w is
different from that used to define the simple maximal arcs
on the neighborhood D of §5. Two different maximal boundary arcs w may intersect in a finite number of saddle
points. At these points the two arcs do not cross.

Any one of these arcs w will be given as the locally
1 - 1 continuous image P(t) of a real interval (0 $\leq$ t $\leq$ 1)
in case w has end points, or of a circle of length 1 with
P(0) = P(1) in case w is a cycle. In the latter case t
can be regarded as the arc length on the circle.

In the case w contacts itself at multiple points, it
will be convenient to introduce a covering w* of w in
which each such multiple point is taken as many times as
it occurs on w. More precisely, one replaces P(t) by the

[*]We suppose P at the level c.

pair $(t, P(t))$. These pairs represent w* in a 1 - 1 continuous manner; taken in this manner w* is the homeomorph
of the interval $(0 \leq t \leq 1)$ or of the unit circle, according as w has or does not have end points.

The set U_c will similarly be replaced by a <u>covering</u>
set U_c^* with the various curves w* as its boundary arcs at
the level c. More precisely, this means that two sectors
S_1 and S_2 of a canonical neighborhood of a saddle point
P_0 at the level c, both of which belong to U_c and which
intersect in P_0, will be regarded as without intersection
when belonging to U_c^*.

It will clarify the following sections to note the
lemma:

<u>LEMMA</u> 8.1. <u>The maximum distance of points on the set</u>
<u>at the level</u> c ± e, <u>from the point set at the level</u> c
<u>tends to zero as</u> e <u>tends to</u> 0.

Were this lemma not true there would exist an infinite sequence of points z_n at levels tending to c as n
became infinite, while z_n tended to a point z_0 on $\overline{G}$ not at
the level c. This would contradict the continuity of U at
z_0. Hence the lemma is proved.

§9.   <u>The Euler characteristic</u> E_c

The Euler characteristic of a 2-complex is defined as

$$a_0 - a_1 + a_2$$

where a_i is the number of i-cells $(i = 0, 1, 2)$ in the
complex. The set U_c may contain a finite number of isolated points of relative minimum of U at the level c. A-
part from these points U_c can be broken up into a finite
set of 2-cells with their bounding 1 and 0-cells. We
shall admit as 2-cells the homeomorph of any convex poly-

gon, regarding the vertices and edges of this image as
0 and 1-cells respectively. Let K_c then be a complex
representing U_c and let E_c be the Euler characteristic
of K_c.

The fact that a complex K_c representing U_c exists
for each c will appear later as we let c increase through
the various critical values of U. In particular it follows from Theorem 7.2 that as c increases through a relative minimum of U, a disjoint 2-cell is added to K_c corresponding to each boundary point at which U assumes the
relative minimum c. It will also appear that a complex
K_c representing U_c can be obtained from any complex K_{c-e}
representing U_{c-e} by adding a finite number of cells,
provided e $>$ 0 is sufficiently small. Or again, if e is
sufficiently small and K_c is given, one can obtain K_{c+e}
from K_c in the same manner.

We shall have occasion to subdivide various 1-cells
of K_c. The resulting change in E_c is null since the addition of a 0-cell at an inner point of a 1-cell is com-
pensated for by the fact that the 1-cell is replaced by
two 1-cells. Note also that a 1-complex representing a
simple closed curve has the same number of 0-cells as
1-cells and so contributes nothing to the Euler characteristic. Accordingly when a 2-cell A_2 is added to K_c on
account of an increase of c through a relative minimum of
U, with A_2 not connected to the other cells, then E_c increases by 1.

As will be seen in the next section, the only changes
in E_c which occur as c increases through the range of
values of U are the increases described in the preceding
paragraph and the changes in the characteristic caused by
changing from U_c^* to U_c when c is the level of a saddle
point. Let K_c^* be a complex representing U_c^* and differing
from K_c in that a 0-cell covering any saddle point P at
the level c appears in K_c^* a number of times equal to the
number m of sectors of a canonical neighborhood of P be-

low c with a vertex at P. To obtain K_c from K_c^* one thus
replaces the 0-cells of K_c^* covering P by a single such
0-cell and modifies the incidence relations accordingly.
The total decrease in passing from E_c^* to E_c on account of
such saddle points equals the sum of the multiplicities
of the saddle points at the level c.

The value of E_c when U_c is empty is 0. The value of
E_c when U_c includes the whole of $\overline{G}$ is 2 - v, where v is
the number of boundary arcs of G. This final value 2 - v
of E_c must equal the algebraic sum of the changes in E_c
as c increases through all values of U. Thus it will appear that

(9.1) $2 - v = m - S - s$

where
  m = the number of points of relative minimum of U
  S = "   "   "   saddle points of U on G
  s = "   "   "   "   "   "   "   "   (B)
counting saddle points according to their multiplicities.
Until §12 is reached, we are assuming that there are no
logarithmic poles. To finish the proof of (9.1) one must
show that the only changes in E_c as c increases arise
from the addition of the disjoint cells corresponding to
the points of relative minimum of U and the changes from
K_c^* to K_c when there are saddle points at the level c.
This is the objective of the next two sections.

§10. <u>Obtaining K_c^* from K_{c-e}</u>

Given c, we impose four conditions on e. The first
condition is that neither U nor -U have critical values
on the interval

                   $c - e \leqq U < c$ $(0 < e)$.

Let w* be any maximal boundary arc of K_c^* at the level c.
There may be several such arcs w* at the level c, but by
convention these arcs have no intersection, although
their projections on K_c may intersect in saddle points at
the level c. Corresponding to each arc w* let $H_w(e)$ be
the subset of points of U_c^* which is connected to w* on
U_c^* among points at which

$$c - e \leq U \leq c.$$

We suppose e so small that no set $H_w(e)$ is connected in
this way to more than one of the arcs w*.

   We also suppose e so small that points of (B) which
are covered by any $H_w(e)$ are saddle[*] points <u>at the level</u>
c in case w* is a cycle; and in case w* is not a cycle,
make up two disjoint U-trajectories T_1 and T_2 on U_c terminating at the end points P_1 and P_2 of w* together with
a subset (possibly empty) of saddle[*] points at the level c.

   The fourth and last condition on e is that e be so
small that the boundary of $H_w(e)$ at the level c - e consist of a single arc. That this condition can be satisfied may be seen with the aid of the construction used in
§7 to establish the isolated character of multiple points.
For that purpose, let P be an arbitrary point of w*, N a
canonical neighborhood of P, and S a sector of N below c
with an arc of w* on its boundary. Let h_1 and h_2 be two
U-trajectories in S emanating from points of w*. If e is
sufficiently small there will be a unique arc in S joining a point of h_1 to a point of h_2 at the level c - e.
Now w* can be banked with a finite number of successively
overlapping sectors S of the above type. Distribute U-
trajectories such as h_1 and h_2 along w* so that successive U-trajectories lie in the same sector S. By the
reasoning of Lemma 7.1, it appears that e can be taken so
small that any arc at the level c - e which meets one of
these U-trajectories meets all of them in the order of the

---

[*]Saddle points of U or of -U.

feet of the U-trajectories on w*. The fourth condition
on e can thus be satisfied.

Let e_1 be a positive constant such that for

(10.1) $0 < e < e_1$

all the conditions on e are satisfied for each boundary
arc w*.

Let g(e) be the boundary of $H_w(e)$ at the level c - e.
The curve g(e) is simple. When not a cycle g(e) has end
points P_1 and P_2 on (B); $H_w(e)$ is then bounded by a Jordan curve formed from g(e), T_1, w*, and T_2. It is accordingly representable as a 2-cell. The vertices of
this 2-cell shall include P_1, P_2, the points of w* which
cover saddle points and any vertices of K_{c-e} which appear
on g(e). The addition of such a 2-cell along a single
arc g(e) causes no change in the Euler characteristic of
K_{c-e}.

In the case g(e) is a cycle, w* with g(e) bounds a
topological annulus on U_c^*. This annulus may be broken
up into 2-cells. Its 0-cells should include the points
of w* covering saddle points together with the 0-cells of
K_{c-e} on g(e). The addition of such an annulus to K_{c-e}
along g(e) makes no change in the Euler characteristic.

Corresponding to each point of relative maximum of U
at the level c there must be added to K_{c-e} to obtain K_c^*
a 2-cell on which

$$c - e \leq U \leq c.$$

This 2-cell is added along a single arc of K_{c-e} and causes
no change in the Euler characteristic.

If there are m_c points of relative minimum of U at
the level c, it is clear that on this account E_{c-e} must
be increased by m_c to obtain E_c^*. It remains to prove the

following lemma:

LEMMA 10.1. The set U_c^* may be obtained as the sum
of U_{c-e}, the points of relative minimum of U at the level
c, the sets $H_w(e)$ corresponding to the various maximal
boundary arcs w* of U_c^* .(e < e_1), and the 2-cells corresponding to the points of maximum at the level c.

Let H be the set of points of U_c^* not given by the
sum in the lemma. Suppose H not empty. On H

(10.2) $c - e < U < c.$

The set H is both closed and open relative to the set
$C = U_c^* - U_{c-e}$ since C - H has this property. Since C
includes all of its limit points not at the level c - e,
H includes a point P at which U has a maximum on H.
Since H is open relative to C and (10.2) holds on H, H is
open relative to $\overline{G}$; hence the point P is a relative maximum of U on $\overline{G}$ as well as on H. Such a maximum is contrary
to the first condition on e. Hence H is empty and the
lemma is true.

We thus have the following theorem:

THEOREM 10.1. If e is a sufficiently small positive
constant

$$E_c^* = E_{c-e} + m_c$$

where m_c is the number of points of relative minimum of U
at the level c.

§11. The variation of E_c with increasing c.

The theorem of the last section needs to be supple-
mented by the following theorem:

THEOREM 11.1. *If* e *is a sufficiently small positive constant*

(11.1) $E_c = E_{c+e}.$

The maximal boundary arcs of U_c used in the preceding section are not adequate for the purpose of proving this theorem. They should be replaced by maximal boundary arcs w+ of U_c with "continuation" through saddle points taken along the boundaries of sectors *above* c rather than below c. As previously, a sector S above c· with vertex at a boundary point P of (B) will be said to be of *boundary type* if there is but one boundary arc of S which tends to P at the level c. An arc w+ which enters a sector S which is above c and is of boundary type shall terminate at the vertex P of this sector.

One then considers the sets $H_{w+}(e)$ for which

(11.2) $c \leqq U \leqq c + e$

and which are connected to the respective arcs w+, restricting e essentially as in the preceding section. As a consequence, these sets $H_{w+}(e)$ will either be representable as distinct 2-cells or topological annuli having a boundary arc at the level c in common with K_c and causing no change in the Euler characteristic when added to K_c. Corresponding to each point of relative minimum P of U at the level c, U_c contains P as an isolated 0-cell, and if e is sufficiently small U_{c+e} includes a 2-cell which forms the closure of a canonical neighborhood of P on (11.2). This change from U_c to U_{c+e} causes no change in the Euler characteristic. Theorem 11.1 follows.

Thus as c increases from a value for which U_c is empty to a value b for which $U_b = \overline{G}$, the only changes in E_c occur when c passes through a critical value c. For e sufficiently small,

$$E_c^* = E_{c-e} + m_c$$

as stated in Theorem 10.1, where m_c is the number of points of relative minimum at the level c. If s_c is the number of saddle points of U at the level c, counting these points according to their multiplicities, then

$$E_c = E_c^* - s_c$$

since U_c is obtained from U_c^* by identifying s_c 0-cells with other 0-cells. Hence

$$E_c = E_{c-e} + m_c - s_c.$$

We have seen in Theorem 11.1 that $E_{c+e} = E_c$ for any sufficiently small e. Hence the total algebraic change in E_c as c increases through all values of c is m - S - s where S, s and m are as defined in §9. We thus have the basic theorem.

THEOREM 11.2. If U has no logarithmic poles, S saddle[*] points on G, s saddle points on (B), and m points of relative minimum, then

$$2 - v = m - S - s$$

where v is the number of boundary curves of G.

---

*Always counting saddle points with their multiplicities.

§12.   The principal theorem under boundary conditions A.

We now include logarithmic poles at interior points of G and begin with the following lemma:

LEMMA 12.1.   If z_0 is a pole of U(x, y) there exists a set of coordinates (u, v) admissibly representing a neighborhood N of z_0 such that the level curves of U in the (u, v) plane are circles with center at z_0 in N.

For simplicity, suppose that $z_0 = 0$. One can refer the neighborhood of the origin to coordinates (x_1, y_1) with $z_1 = x_1 + iy_1$ such that U becomes a function

$$(12.1) \qquad k \log |z_1| + kR[F(z_1)], \quad [R = \text{real part}]$$

where k is a real non-null constant and F(z) is analytic in z_1 at $z_1 = 0$. The function (12.1) can be written in the form

$$(12.2) \qquad k \log \left| z_1 \, e^{F(z_1)} \right|.$$

The transformation

$$w = u + iv = z_1 \, e^{F(z_1)}$$

is 1 - 1 and conformal neighboring the origin. By virtue of this transformation the given pseudo-harmonic function takes the form

$$(12.3) \qquad k \log |w|$$

in a sufficiently small neighborhood of w = 0, and the level lines are the circles w = constant.

The principal theorem may be stated as follows:

THEOREM 12.1. Let U be pseudo-harmonic except for logarithmic poles on a limited region G bounded by v Jordan curves, with U continuous at points of the boundary (B) of G. If the function defined by U on (B) has at most a finite set of points of relative extremum, then

(12.4) $M + m - S - s = 2 - v$

where M is the number of logarithmic poles of U on G, m the number of points of relative minimum of U, while S and s are respectively the number of interior and boundary saddle points of U each counted with its multiplicity.

To prove the theorem we shall replace G by a region G_1 from which each pole P of U is excluded by a new boundary curve B(P). To define B(P) refer a neighborhood of P coordinates (u, v) as in Lemma 12.1 so that the level curves neighboring P are circles with center at P. In this neighborhood take B(P) as a small circle containing P with a center not at P. The poles P may be excluded in this way by new boundaries which do not intersect each other or the original boundaries. Let U_1 be the function defined by U on G_1. The numbers m and s belonging to U on G will be replaced by numbers m_1 and s_1 belonging to U_1 on G_1.

It will appear that

(12.5) $m - s = m_1 - s_1.$

Given the pole P, two cases must be distinguished according as |w| of Lemma 12.1 increases or decreases with increasing U near P. If |w| increases with U, it is seen that U_1 has a point of minimum relative to $\overline{G}_1$ on the boundary B(P), and a saddle point relative to $\overline{G}_1$ of

multiplicity 1 at the point on B(P) at which |w| is a
maximum. The level lines |w| = constant make this clear.
The contribution to m_1 - s_1 from such a pole is thus
1 - 1 or 0. It is similarly seen that when |w| decreases
with increasing U neighboring P, U_1 has no critical
points on B(P). Hence (12.5) holds as stated.

The number S is the same for U_1 on G as for U on G.
If v_1 is the number of boundaries of G_1

$$(12.6) \qquad 2 - v_1 = m_1 - s_1 - S = m - s - S$$

in accordance with the theorem of §11. The theorem follows from (12.6) and the relation v_1 = v + M.

COROLLARY. Under the hypotheses of the theorem

$$S - M \leqq v - 2 + m.$$

In particular, when there are no poles and just one
boundary

$$(12.7) \qquad\qquad S \leqq m - 1.$$

Examples. We shall show that equality may occur in
(12.7) while at the other extreme S can be zero and m an
arbitrary positive integer.

To that end, let

$$U \equiv R(z^m) \equiv r^m \cos m\theta.$$

Taking U on the disc $r \leqq 1$, the boundary values cos $m\theta$
have m points of minimum at the level -1, and m points of
maximum at the level 1. These points of maximum of
cos $m\theta$ give maxima of U. The origin is a saddle point of

multiplicity m - 1. Thus

$$S = m - 1.$$

In our second example we shall suppose that $U = y$. The domain $\overline{G}$ of definition of U shall be of the form

$$f_1(x) \leqq y \leqq f_2(x), \qquad\qquad 0 \leqq x \leqq (n + 1)\pi$$

with

$$f_1(x) = -\sin^2 x$$
$$f_2(x) = \sin^2 \frac{x}{n+1} \ .$$

On the boundary y assumes its absolute maximum at the point

$$x_0 = \frac{(n+1)}{2}\pi \qquad\qquad y_0 = f_2(x_0) = 1.$$

On the curve $y = f_1(x)$, y has $n + 1$ relative minima and n relative maxima on the open interval $[0, (n+1)\pi]$. The level curves of U are the lines $y = $ constant, and it is seen that each of these maxima of $y = f_1(x)$ give saddle points of y relative to $\overline{G}$. Thus

$$S = 0, \qquad m = n + 1, \qquad s = n, \qquad V = 1.$$

A particular consequence of the theorem is obtained by comparing $-U$ with U. If S, M, m and s refer to U, then S and M will be the same for $-U$. On comparing (12.4) for U with (12.4) for $-U$ it is seen that

$$\overline{s} - \overline{m} = s - m$$

where "-" refers to -U. In general, the boundary saddle
points of U and -U will differ as sets and in individual
multiplicities.

The function U must either have a logarithmic pole
on which U becomes negatively infinite or else $m \geq 1$.
The number $v \geq 1$. The integers M, S, s are positive or
zero. The relation in Theorem 12.1 must also be satisfied. There are no other relations between these integers
as may be shown by constructing a harmonic function which
realizes an arbitrary set of integers satisfying the
above relations.

§13. The case of constant boundary values

In many important applications the harmonic function
under consideration is constant on one or more of the
boundaries, as for example in the case of a Green's function. To modify our previous results to take care of
such a case the fundamental relation (12.4) will be written in the form

$$(13.1) \qquad\qquad M - S = 2 - v + I$$

and I termed the boundary index. Under Boundary Conditions A, I has been evaluated as s - m. More explicitly,

$$I = \sum (s_i - m_i) \qquad\qquad (i = 1, \ldots, v)$$

where m_i is the number of points of minimum of U on B_i
and s_i the number of saddle points of U on B_i. One can
then properly term $s_i - m_i$ the contribution I_i of B_i to I.
We continue with the evaluation of I in the following
theorem.

THEOREM 13.1. If Boundary Conditions A are satisfied
on a subset of the boundaries (B) and if on each of the

remaining boundaries C_i U is a constant relative extremum, then I in (13.1) is the sum of the contributions
$s_j - m_j$ of the boundaries which satisfy Conditions A,
with no contribution from the boundaries C_i.

To establish the theorem, we refer to the variation
of the Euler characteristic E_c with c as described in §11,
and show that when c varies through the level c_i of C_i
there is no change in E_c on account of C_i.

Suppose in particular that c_i is a relative maximum
of U. Let N_i be a neighborhood of C_i relative to G in
the form of a topological annulus so near to C_i that $\bar{N}_i$
includes no boundaries of G other than C_i and no poles of
U. If e is a sufficiently small positive constant the
locus $U = c_i - e$ includes a closed curve C_i^e on N_i. The
curve C_i^e must be non-bounding on N_i since U has no extrema on N_i. Hence C_i^e must be simple, and together with C_i
bound a topological annulus R_i on $\bar{N}_i$. In passing from
the complex K_{c-e} to K_c as in §11, the addition of R_i to
K_{c-e} will make no change in the Euler characteristic. In
case c_i is a relative minimum of U the analysis is similar, on noting that the Euler characteristic of an annulus
is zero.

The theorem follows.

Recall that the Green's function for G has a constant maximum value zero on each boundary and a logarithmic pole at which U becomes negatively infinite at a prescribed point of G. We thus have the following consequence of Theorem 13.1.

COROLLARY 13.1.   The Green's function for G has
v - 1 saddle points, counting these saddle points with
their multiplicities.

In this application M = 1, and I = 0, so that
$S = M + v - 2 - I = v - 1$ as stated.

Boundaries C on which U is a constant c but not

necessarily a relative extremum, make a contribution to I
which can be evaluated. One supposes in this case that
U is pseudo-harmonic on a neighborhood of C relative to
the (x, y) plane. There will be at most a finite set of
points Q on C to which arcs at the level c tend from G.
The number 2σ of such arcs will be even, and the integer
σ will be termed the level index of C relative to G. In
counting the number of arcs tending to C from G, arcs
tending to different points of C are counted as different,
and arcs tending to the same point Q of C are counted as
different if they appear as different radial arcs on a
canonical neighborhood of Q.

We need a description of the level arcs on G neighboring C. We shall represent a neighborhood of C relative to $\overline{G}$ as the closed join of a sequence of sectors
neighboring C and between successive level arcs tending
to C. These sectors will be of two types; a V-type of
sector lying between two level arcs which tend to the
same point of C, and an R-type of sector between two
level arcs h and h' tending to different points Q and Q'
of C. A V-type of sector and its level arcs are readily
described; for the composition of f with an appropriate
homeomorphism of a neighborhood of Q leads to a harmonic
function with saddle point at Q. Thus the V-types of
sectors which arise from harmonic function with a saddle
point at Q can serve as models of V-type sectors in general.

As a model of an R-type of sector, we shall use a
rectangle H in which the straight lines parallel to the
base of H will serve as representatives of the level arcs
of U. More explicitly, let h and h' be two level arcs of
U on G tending to points Q and Q' of C between which lies
an arc k of C. Let $\overline{G}$ be cut along h and h' to form a
domain K. We suppose h and h' so chosen that there are
no level arcs of U other than h and h' on K tending to k.
There exists an arbitrarily small neighborhood N of k

relative to K, with $\overline{N}$ the homeomorph of H and with level arcs of U on $\overline{N}$ corresponding to parallels to the base of H on H. The base of H corresponds to k preceded and followed respectively by sub-arcs of h and h'. The sides of H correspond to U-trajectories emanating from points of h and h' respectively.

The principal theorem is as follows.

THEOREM 13.2. Let U be pseudo-harmonic on a region which includes $\overline{G}$ in its interior, except for logarithmic poles on G. If Boundary Conditions A are satisfied on a subset of the boundaries and if U is a constant c_i on each remaining boundary C_i, then the boundary index I of (13.1) is

$$(13.2) \qquad\qquad \sum(s_j - m_j) + \sum \sigma_i$$

when $s_j - m_j$ is summed over the boundaries on which Conditions A are satisfied, and the level indices σ_i are summed over the boundaries C_i on which U is constant.

Each boundary C_i is replaced by a nearby simple closed curve C_i' such that Boundary Conditions A are satisfied on C_i' relative to the modified domain $\overline{G}'$. To that end C_i' is drawn so that U on C_i' (denoted by U_i) increases or decreases on C_i' without exception except for one point of relative maximum of U_i on C_i' in each sector above c_i, and one point of relative minimum of U_i on C_i' in each sector below c_i. The relation of C_i' to the level arcs of U, as shown by the respective sector models, proves that the points of relative maximum of U_i on C_i' are saddle points of U relative to $\overline{G}'$ of multiplicity 1, while the points of relative minimum of U_i on C_i' are ordinary.

The sectors above c_i and below c_i alternate, and the number of sectors above c_i is σ_i. The theorem follows from the results established earlier under Boundary Conditions A.

An example.   Consider the harmonic function

$$U(x, y) = \log \left| \frac{1 + 3z^2}{z(3 + z^2)} \right| \qquad (|z| \leq 1).$$

It will be seen in §21 that M = 3 and S = 0 on $|z| < 1$
while U = 0 on $|z| = 1$.   The arcs at the level 0 tending
to the circle $|z| = 1$ from the region $|z| < 1$, include two
arcs tending to z = 1 and two arcs tending to z = -1.
Thus $\sigma = 2 = $ I.   The relation

$$M - S = 1 + I = 3$$

is satisfied.

CHAPTER II

DIFFERENTIABLE BOUNDARY VALUES

§14. Boundary conditions A, B, C.

These conditions are defined as follows.

Boundary Conditions A require that the boundary values of U have a finite number of points of extremum. The boundaries admitted are arbitrary Jordan curves.

Boundary Conditions B* require that a neighborhood N of (B) in the (x, y) plane exist over which U can be extended in definition so as to be of class C' and ordinary. The boundaries are supposed regular.

Boundary Conditions C* require that Conditions A and B be satisfied.

Recall that a function U(x, y) is of class C' on a region R if U_x and U_y exist on R and are continuous. A function of class C' is ordinary on R if $U_x^2 + U_y^2$ does not vanish on R. An arc is termed regular if it is representable in the form

$$x = x(t); \qquad y = y(t) \qquad (a \leqq y \leqq b)$$

where x(t) and y(t) are of class C' and

---

*Conditions B or C can be replaced by the conditions that some neighborhood N of (B) possess a coordinate system with respect to which U satisfies Conditions B or C and in terms of which the boundary is regular. We refer to these new conditions as Generalized Conditions B or C. The only coordinate systems which will be admitted are those obtained by a sense-preserving transformation from (x, y).

$$x'^2(t) + y'^2(t) \neq 0.$$

A closed curve is termed regular if each sub-arc is regular. One can always take the arc length as parameter on
a regular curve.

Under boundary conditions B, U has a non-null gradient g at each point of its boundary. This gradient is
the vector whose components are U_x, U_y. A point P on the
boundary will be termed entrant if g enters G, or more
precisely if g may be obtained from the positive tangent
to (B) at P by rotating the tangent through a positive
angle less than π . Recall that the exterior boundary of
G is ordinarily so sensed that its order with respect to
a point within it is 1, while the other boundaries of G
are sensed so as to have an order of -1 with respect to
points within. A boundary point P at which the gradient
does not lie on the tangent and which is not entrant will
be termed emergent.

At points P at which U is of class C' and ordinary
the gradient is orthogonal to the level curve through P,
as one sees from the relation

$$U_x dx + U_y dy = 0$$

which holds along the level curve. Under boundary conditions B or C we shall ordinarily represent the boundary
values as a function of the arc length s on the boundary
considered. With boundary values so represented a necessary and sufficient condition that the gradient be normal
to the boundary at a point P is that the s-derivative of
the boundary value function be null at P. For this condition means that the tangential component of the gradient vanishes. Each boundary point P at which the gradient
is not normal to the boundary is ordinary relative to $\overline{G}$ in
the sense of the earlier sections. For on the extended

neighborhood of P there is a single regular level arc
through P making a non-zero angle with the tangent to the
boundary at P so that there is but one level arc tending
to P on G. P is then clearly neither a point of extremum
of U nor a saddle point.

There remain the boundary points at which the s-derivative of the boundary values U^1 vanish. We shall
treat such points first under Boundary Conditions C, and
shall establish the following lemma:

LEMMA 14.1. Under Boundary Conditions C,[*] a minimum
of U occurs only at an entrant relative minimum of U^1,
and a saddle point only at an entrant relative maximum of
U^1. Each saddle point has the multiplicity 1.

To prove the lemma we enumerate the different cases
against the type of point (ordinary, minimum, saddle
point relative to $\overline{G}$).

	The Case	The type of point, rel. $\overline{G}$
(a)	An entrant maximum of U^1	A saddle point
(b)	An emergent maximum of U^1	A maximum of U
(c)	An entrant minimum of U^1	A minimum of U
(d)	An emergent minimum of U^1	An ordinary point
(e)	An ordinary point of U^1	An ordinary point

The key to the proof lies in the fact that in the
extended neighborhood of the boundary point P there is
just one arc at the level c of P passing through P. This
level arc intersects the boundary B_1 only at P neighboring P. On G neighboring P there are accordingly two, one,
or no arcs at the level c terminating at P. The only possible saddle points have the multiplicity 1.

Case (a). The two boundary arcs tending to P neighboring P are below c while the inner normal is above c

___

*The extension of this lemma to Generalized Conditions C
is obvious. The notion of an entrant or emergent gradient is then relative to the coordinate system used.

neighboring P. Hence there are two sectors below c with vertex at P. Thus P is a saddle point relative to $\overline{G}$.

Case (b). The function U comes under Case (a) relative to the complement of G. Hence there is but one sector on G neighboring P, and this sector is below c.

Case (c). The function -U comes under (b) relative to G, so that -U has a maximum at P and U a minimum.

Case (d). The function -U comes under Case (a) relative to G, so that relative to U there is just one sector below c on G with vertex at P.

Case (e). Neighboring P one boundary arc is below c and one above c. There must then be an even number of sectors on G with vertex at P. Since this number of sectors is at most three and not zero, it must be two. Thus P is ordinary.

The lemma follows and leads to a new theorem:

THEOREM 14.1. <u>Under</u> <u>Boundary</u> <u>Condition</u> C <u>the</u> <u>numbers</u> m <u>and</u> s <u>in the</u> <u>relation</u>

$$M - S = 2 - v + s - m$$

<u>may</u> <u>be</u> <u>evaluated</u> <u>as</u> <u>the</u> <u>number</u> <u>of</u> <u>entrant</u> <u>points</u> <u>of</u> <u>minimum</u> <u>and</u> <u>maximum</u> <u>respectively</u> <u>of</u> <u>the</u> <u>boundary</u> <u>values</u>.

This theorem holds equally well for Generalized Conditions C with "entrant" interpreted in terms of the coordinate system used.

<u>Computational</u> <u>use</u> <u>of</u> <u>Theorem</u> 14.1. A particularly simple application of Theorem 14.1 is the determination of the number and position of the zeros of F'(z) when F(z) is analytic with at most logarithmic singularities. The following example will illustrate the procedure and the boundary data necessary for the calculation. Let

$$F(z) = \frac{z^4}{4} + z^2 - iz + \log z.$$

It is required to find the number of zeros of $F'(z)$ for which $|z| < 1$.

One uses polar coordinates (r, θ) and sets

$$R[F(z)] = u(r, \theta).$$

Then the boundary values of u may be found from the equation

$$u = \frac{r^4 \cos 4\theta}{4} + r^2 \cos 2\theta + r \sin \theta + \log r$$

on setting $r = 1$. On plotting $u(1, \theta)$ as indicated in Fig. 2, it is found that $u_\theta(1, \theta) = 0$ at six values of θ:

$$0 < a_1 < a_2 < \ldots < a_6 < 2\pi.$$

To find out which, if any, of these values of θ is extrant, one evaluates

$$u_r = r^3 \cos 4\theta + 2r \cos 2\theta + \sin \theta + \frac{1}{r},$$

when $r = 1$. It is found that $u_r(1, a_1) < 0$ only at a_6 so that the boundary point $(r, \theta) = (1, a_6)$ alone is entrant. Since this point affords a minimum to u it follows that in Theorem 14.1, $m = 1$, $s = 0$ and $M = 1$. The relation

$$M - S = 1 + s - m$$

yields the result $S = 1$. Hence $F'(z)$ vanishes just once for $r < 1$.

A computation of the above character takes less than fifteen minutes   On using other circles $r = r_0$ one can locate the value of r representing a root of $F' = 0$ with any required degree of accuracy.

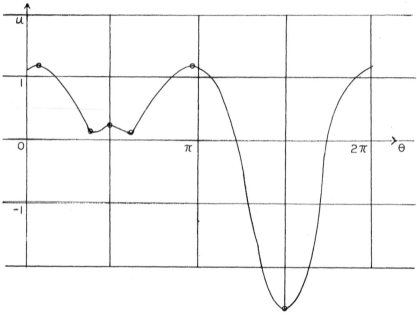

Figure 2.

§15.   Boundary conditions B

Our fundamental relation has been written in the form

(15.1)   M - S = 2 - v + I

where I has been evaluated under Boundary Conditions A as the difference s - m between the number of boundary saddle points and the number of relative minima of U, and under Boundary Conditions C as the difference s - m between the number of entrant points of relative maximum and minimum of the boundary values. We term I, as de-

fined by (15.1), the boundary index of U, and shall find
a method for evaluating I under Boundary Conditions B as
well as under Conditions A.

In the treatment of this case we shall need the notion of a line element defined by a point (x, y) and pair
of direction cosines (a, b),

$$a^2 + b^2 = 1.$$

The line element shall be a point

(15.2)   (x, y, a, b)

in Cartesian 4-space, and the distance between two line
elements shall be the ordinary distance between the
points in 4-space representing the line elements.

We shall need the Fréchet distance between two
curves. See Fréchet (1). We consider the case of an
arc* g_1

(15.3) $x_1(t), y_1(t)$ $(a \leq t \leq b)$

given as the continuous but not necessarily 1 - 1 image
of a line segment. Let g_2 be a second arc with a parameter u ranging on an interval (c, d). Let T be any homeomorphism between the intervals (a, b) and (c, d) with a
corresponding to c, b, to d. Under T there will be a
maximum distance D(T) between points which correspond
under T. The Fréchet distance between the curves g_1 and
g_2 is taken as the greatest lower bound of F(T) as T
ranges over all homeomorphisms between the intervals
(a, b) and (c, d). This number will be called the
0-order distance as distinguished from a 1st-order dis-

---

*The case of a closed curve is similar

<u>tance</u> to be defined next.

Suppose that g_1 and g_2 are regular so that t and u may be taken as arc lengths. Then

(15.3.1) $[x_1(t),\ y_1(t),\ x_1'(t),\ (y_1'(t)]$

and

(15.3.2) $[x_2(u),\ y_2(u),\ x_2'(u),\ y_2'(u)]$

become curves in 4-space composed of line elements tangent, respectively, to g_1 and g_2. The Fréchet distance in 4-space between the curves (15.3.1) and (15.3.2) will be called the 1<u>st order distance</u> between g_1 and g_2.

We shall consider transformations of coordinates neighboring a point $(x_0,\ y_0)$ which admit the form

(15.4)
$$u = u(x,\ y)$$
$$v = v(x,\ y)\ ,$$

where $u(x,\ y)$ and $v(x,\ y)$ are of class C' on a neighborhood of $(x_0,\ y_0)$ and the Jacobian

$$J = u_x u_y - u_y v_x$$

is not zero at $(x_0,\ y_0)$. One restricts the transformation to a circular neighborhood N of $(x_0,\ y_0)$ on whose closure the transformation is 1 - 1 and the Jacobian is not zero.

Let s be the arc length in the (x, y) plane and s_1 the arc length in the (u, v) plane. If dx and dy are differentials along a regular curve g in the (x, y) plane, then along the image of g

(15.5) $ds_1^2 = Edx^2 + 2Fdxdy + Gdy^2$,

where

$$E = u_x^2 + v_x^2 \qquad F = u_x u_y + v_x v_y \qquad G = v_x^2 + v_y^2.$$

Since

$$EG - F^2 = J^2 \neq 0$$

the right member of (15.5) is positive definite. It is
bounded from zero if (dx, dy) are direction cosines,
that is, if $dx^2 + dy^2 = 1$. A line element of the type

(15.6)   (x, y, dx, dy) $(dx^2 + dy^2 = 1)$

emanating from a point (x, y) on N is transformed into a
line element

(15.7) $(u, v, \dfrac{du}{ds_1} \dfrac{dv}{ds_1})$

which varies continuously with the element (15.6) for
(x, y) on N. Here du, dv, ds_1^2 are to be regarded as
polynomials in the direction cosines (dx, dy) with coefficients dependent on (x, y). The polynomial ds_1^2 is
bounded from zero as a function of the elements (15.6)
for (x, y) on N.

Considering Boundary Conditions B, let $P = (x_0, y_0)$
be an arbitrary point of the boundary of G. We make a
transformation T(P) of a neighborhood of P of the form

(15.8) $u = U(x, y) \qquad v = h(x - x_0) + k(y - y_0)$, where

$$h = -U_y, \qquad k = U_x, \qquad (\text{at } (x_0, y_0)).$$

This choice of h and k has as a consequence that

$$J = U_x^2 + U_y^2 \neq 0 \qquad (\text{at } x_0, y_0).$$

Thus J is the square of the length of the gradient of U
at P. By hypothesis U is ordinary at each boundary point
so that $J \neq 0$ at P.

The transformations T(P) thus form a family of transformations varying continuously with the point P on (B).
At the initial point P of each transformation $J \neq 0$. The
domain (B) of P is closed and J varies continuously with
(x, y) and P. Therefore there exists a constant r_0 which
is so small that on the closure of the circular neighborhood N(P) of P of radius r_0 the transformation T(P) is
1 - 1, with a Jacobian which is both bounded and bounded
from zero for (x, y) on N(P) and P on (B), while both
T(P) and its inverse transform line elements continuously, uniformly with respect to the parameter P on (B).

Let B' be a particular boundary curve in the set (B).
We shall prove the following lemma:

LEMMA 15.1. Under Boundary Conditions B there exists
a regular curve B" within an arbitrarily small first order Fréchet distance e from a given boundary curve B' of
the set (B) such that Boundary Conditions A are satisfied
on B".

Let r* be any positive constant less than the radius
r_0 of the neighborhoods N(P). Let

$$(15.9) \qquad\qquad P_1, \ldots, P_n$$

be a circular sequence of points of B' such that the arc
of B' from P_i to its successor has a length at most r*.
If $U(P) = U(Q)$, for any two successive points P, Q of the
sequence (15.9), let P or Q be slightly displaced so that
$U(P) \neq U(Q)$. We then join P_i to its successor by an arc
h_i which is straight in the space (u, v) defined by the
transformation $T(P_i)$. The level curves of U in the space
(u, v) are the parallel straight lines on which u is
constant; therefore on each arc h_i U is strictly monotone.
The arcs h_i of the circular sequence combine to define a
curve B*. A corner in B* at a point P_i will be rounded
off by a small circular arc in the coordinate system
(u, v) defined by $T(P_i)$; this is particularly simple because the arc h_i is straight in the coordinate system
(u, v). The resulting regular curve will be denoted by
B".

The curves B' and B* can be put into 1 - 1 continuous correspondence by making the end points of h_i correspond to the end points of the arc of B' which h_i replaces,
completing this correspondence over h_i by linear interpolation with respect to arc length on B' and B*. A
homeomorphism between B* and B" can be similarly defined
resulting in a homeomorphism between B' and B".

The curve B" depends on the choice of r*, upon the
preliminary displacement (if any) of the points P_i from
positions on B', and on the rounding off of the corners
of B*. If r* is taken sufficiently small, and the points
P_i are displaced through correspondingly small distances,
then the resulting curve B" will lie within the prescribed 1st-order Fréchet distance of B'. At any rate, this
is clear for corresponding sub-arcs of B' and B" in a
local coordinate system (u, v); and it then follows in
the plane (x, y) by virtue of the way in which line elements are transformed under the transformations $T(P_i)$.

§16.   The vector index J of the boundary values

We are assuming that Boundary Conditions B hold. A vector index J_i of U will be defined on each boundary B_i. The vector index J of the whole boundary shall be the sum of its indices J_i.

By an entrant covering* W of B_i will be meant any finite set of closed sub-arcs of B_i which include no points of B_i at which the gradient g of U is normally emergent, and whose interiors are disjoint and include all points of B_i at which g is normally entrant.

The whole of a boundary B_i may belong to W. The set W may be empty if, for example, there are no entrant normal gradients. Under Conditions B the feet of entrant normal gradients on B_i form a closed set at a positive distance from the set of feet of emergent normal gradients. It follows that an entrant covering W of B_i always exists under Conditions B.

Let g be the gradient of U at a point s of B_i. The vector projection of g onto the tangent to B_i at s will be denoted by g_s. Let h be any arc of an entrant covering W. An end point s of h will be termed tangentially entrant relative to h if g_s is directed toward the interior of h; otherwise s will be termed tangentially emergent relative to h. It should be noted that under Conditions B, g_s is never null at an end point of an arc h of W, since such an end point never bears a gradient which is normal to B_i.

The vector index** J_i of U on B_i will be taken as the number of tangentially entrant end points of arcs of W, minus the number of arcs of W with* end points. If an arc of W is identical with B_i, J_i will be taken as zero.

---

*To take care of the special case in which every gradient is entrant on a boundary B_i we must admit the whole of B_i as a possible arc of W.

**A generalization of this vector index termed the "alternating characteristic" has been defined by Morse for n-dimensional vector fields. See Morse, (3).

A first lemma follows:

LEMMA 16.1. The vector index J_1 of U on B_1 is independent of the choice of the entrant covering of B_1.

Corresponding to any two entrant coverings W_1 and W_2 of B_1 let W be an entrant covering which coincides with $W_1 + W_2$ as a point set, and whose set of 0-cells (i. e., arc end points) is the sum of the 0-cells of W_1 and W_2. Then W_1 or W_2 can be obtained from W by a finite sequence of steps of the following sort:

1) Removing an arc of W which bears no entrant normal gradient.

2) Removing two coincident end points as end points.

Operation (1) does not alter the vector index of B_1 since one end point (if the arc is not B_1) of the arc removed must be tangentially entrant and the other end point tangentially emergent. The vector index of the arc removed is thus $1 - 1 = 0$. The removal of B_1 as an arc of W causes no change in J_1. Operation (2) does not alter the vector index of B_1 since the end point which is removed must be tangentially entrant relative to one abutting sub-arc and tangentially emergent relative to the other.

This completes the proof.

LEMMA 16.2. The vector index J_1 of a boundary arc B_1 equals the vector index of any simple regular curve B_1' which lies within a sufficiently small positive 1st order Fréchet distance e of B_1 and replaces B_1 as a boundary curve.

Let T be a homeomorphism between B_1 and B_1' such that corresponding elements have a distance at most 2e. If e is sufficiently small, the image on B_1' of an entrant covering of B_1 will define an entrant covering of B_1' such that corresponding end points of corresponding arcs h and

and h' of the covering will both be tangentially entrant
or both tangential emergent relative to h and h' respectively. The lemma now follows from the definition of the
vector index of a boundary curve.

LEMMA 16.3. Under Boundary Conditions[*] C,

$$s_i - m_i = J_i$$

where J_i is the vector index of B_i, s_i is the number of
entrant points of relative maximum of U_i, and m_i is the
number of entrant points of relative minimum of U_i.

Let h be an arc of an entrant covering of B_i. If h
has end points there are three cases:
Case (a). Both end points of h are tangentially entrant relative to h.
Case (b). Both end points of h are tangentially e-
mergent relative to h.
Case (c). One end point of h is tangentially entrant and the other tangentially emergent relative to h.
In case (a) the contribution to J_i is 2 on account
of h's end points, and -1 on account of h. The net contribution is 1. In Case (a), U_i increases as h is enter-
ed from either end. Hence the number of points of relative maximum of U_i on h exceeds the number of points of
relative minimum by 1. These extremum points are all entrant. Hence the contribution of h to $s_i - m_i$ is 1, or
equal to the contribution of h to J_i. In Case (b), h
contributes -1 both to $s_i - m_i$ and to J_i. In Case (c),
h contributes 0 both to $s_i - m_i$ and to J_i.
In the special case in which h coincides with B_i,
every extremum point of U_i is entrant so that $s_i - m_i = J_i = 0$.
The principal theorem of this section can now be

[*]Or the Generalized Boundary Conditions C.

proved.

THEOREM 16.1. If a neighborhood of the boundary (B)
admits coordinates (u', v') in terms of which Boundary
Conditions B are satisfied,and if J is the vector index
of the boundary as determined in terms of these coordinates,then

$$M - S = 2 - v + J.$$

Replace each boundary curve B_i by a boundary curve
B_i' which lies within so small a first order Fréchet distance[*] of B_i, that the vector index[*] J_i of B_i equals the
vector index[*] of B_i'. In accordance with Lemma 15.1 one
can take B_i' so that Boundary Conditions[*] C are satisfied
on B_i'. One takes B_i' so near B_i that all of the poles and
saddle points of U lie within the region bounded by the
curves B_i'. It follows from Lemma 16.3 that on B_i'

(16.1) $s_i - m_i = J_i$ $(i = 1, 2, \ldots, n)$.

But according to Theorem 14.1

(16.2) $M - S = 2 - v + \sum_{i=1}^{n} (s_i - m_i)$.

The theorem follows from (16.1) and (16.2).
    An example. Consider the function

$$U(x, y) = x^2 - y^2$$

on the disc $|z| \leqq 1$. The boundary points $(\pm 1, 0)$ bear
emergent normal gradients, and the boundary points

---

[*]Relative to the space of the coordinates (u', v').

(0, $\pm$1) bear entrant normal gradients. We take an entrant covering of the boundary which includes all boundary points at which $y^2 \geq x^2$, employing two entrant arcs.
Their end points are tangentially emergent, so that the
vector index of the boundary is -2. Thus

$$-S = 2 - v + J = 1 - 2 = -1.$$

§17.   The vector index J as the degree
of a map on a circle

We shall prove the following theorem:

THEOREM 17.1. If B_1 is a regular boundary in a
neighborhood of which Boundary Conditions B are satisfied,
then as a point s traces B_1 in its positive sense any continuous branch of the angle $\theta(s)$ from the interior normal
at s to the gradient at s, increases by $2\pi J_1$, where J_1 is
the vector index of B_1.

Remark. This variation of $\theta(s)$ is called the degree
of the map on the unit circle defined by $\theta(s)$. It has
not been used to define J for several reasons. First, as
has been seen in the preceding section, J defined as a
vector index is immediately connected with the distribu-
tion of extremum points of the boundary values and the
earlier evaluations of J. Secondly, the generalization
of the vector index to n-dimensional vector distributions
cannot be readily reduced as in Theorem 17.1 to the degree of a map. Finally, what is most significant, the
vector index as defined in the preceding section can be
defined in related terms when the normal to B_1 and the
gradient fail to exist. The hypothesis that U is ordinary
is replaced by the hypothesis that the level curves as
extended over the boundary are without self intersections
near the boundary. The concepts of entrant and emergent
gradient can be usefully modified in terms of increasing
or decreasing U or U_1 with appropriate hypotheses limit-

ing the boundary. The generalized vector index then becomes a kind of integral.

   Proof of the theorem. Neither the degree of the map
defined by $\theta(s)$ nor J_i will be changed if B_i is replaced,
as in §15, by a regular curve B_i' which has a sufficiently
small 1st order Fréchet distance from B_i and on which Conditions A as well as B are satisfied. As constructed in
§15, B_i' is such that on B_i' the gradient coincides with
the interior normal only at entrant points of extremum of
$U_i(s)$ on B_i', and at these points an appropriate branch of
$\theta(s)$ changes sign. We can choose s' = 0 so that $\theta(0) \neq 0$,
mod 2π. As s increases $\theta(s) = 0$, mod 2π, at only a finite
number of points of B_i'. The theorem depends on the following lemma:

   LEMMA 17.1. If k is an arbitrary arc of an entrant
covering of B_i' the contribution of k to J_i equals the
number of points of k at which $\theta(s)$ increases through 0,
minus the number of points of k at which $\theta(s)$ decreases
through 0, as k is traversed in its positive sense, taking
$\theta(s)$ mod 2π on the interval $-\pi < \theta \leq \pi$.

   If $k = B_i'$ the contribution to J_i is zero and the
lemma is clearly true. Apart from this case the lemma is
a consequence of the following statements:
1) As s traverses k, $\theta(s)$ never equals π, mod 2π.
2) At an initial point of k which is tangentially entrant, $\theta(s)$ lies between 0 and $-\pi$, mod 2π.
3) At a terminal point of k which is tangentially entrant, $\theta(s)$ lies between 0 and π, mod 2π.
4) At an initial point of k which is tangentially emergent, $\theta(s)$ lies between 0 and π, mod 2π.
5) At a terminal point of k which is tangentially emergent, $\theta(s)$ lies between 0 and $-\pi$, mod 2π.
   The cases which can happen are [(2) (3)] [(2) (5)]
[(3) (4)] [(4) (5)].

If, for example, (2) and (3) occur, θ(s) must increase through 0, mod 2π, as s increases over k. On the
other hand, k and its end points contribute 2 - 1 = 1 to
J_1 in this case. Thus the lemma holds in this case.

The other cases are similar. The lemma and the theorem follow.

Let g: u = u(s), v = v(s), be a regular, sensed
closed curve in the (u, v) plane, not intersecting the
origin. Let s be the arc length along g with $0 \leq s \leq \sigma$.
Let N represent a neighborhood of g on g's positive side.
Recall that the positive side of g includes the points
into which the inner normal projects. This normal is obtained on rotating the positive tangent through 90 de-
grees. We do not suppose that g is simple. When g has
multiple points it is clear that one must regard N as a
kind of Riemann ribbon following along g. For this purpose points on g and the corresponding points on N will
be regarded as distinct if associated with different parameter values s with $0 \leq s \leq \sigma$. We shall consider the
function

$$u^2 + v^2 = U(u, v).$$

Let J be the vector index of g relative to the function
U(u, v) on $\overline{N}$, with J defined by an entrant covering of g.

LEMMA 17.2. Let N be a neighborhood of the regular,
closed curve g on the positive* side of g. If J is the
vector index of $u^2 + v^2$ evaluated on g relative to $\overline{N}$, q
the order of g relative to the origin, and p the angular**
order of g, then J = q - p.

---

*Attention is called to the fact that an interior trans-
formation f which is locally 1 - 1 neighboring (B), carries the positive side of a boundary B_1 into the positive
side of any regular image g_1 of B_1. The sense-preserving
character of f enters in this way.
** See §18 for definition.

The proof of Theorem 17.1 applies. The direction of the gradient of $u^2 + v^2$ is that of the radius vector from the origin to g, so that the gradient on g makes q revolutions as g is traversed. The tangent to g makes the same number of revolutions as the interior (or exterior normal). The lemma follows from Theorem 17.1 extended to the case at hand.

CHAPTER III

INTERIOR TRANSFORMATIONS

§18. Locally simple curves

The curves presently to be admitted as boundary
images are continuous and locally simple images g

(18.1) $x(t)$, $y(t)$

of a circle with angular parameter t. The condition of
local simplicity implies that there exists a positive
constant e so small that an arc of g on which $|\Delta t| < e$ is
simple. Let d(e) be the minimum diameter of the set of
sub-arcs of g on which $|\Delta t| \geq e$. It is clear that d(e)
is positive. For any sub-arc h of g whose diameter is
less than d(e) it follows that $|\Delta t| < e$ so that h is
simple. Any constant e_1 such that each sub-arc of g of
diameter less than e_1 is simple will be called a norm of
local simplicity of g. Any set of locally simple curves
admitting the same norm e_1 will be termed uniformly
locally simple.

It is convenient to set

$$z(t) = x(t) + iy(t).$$

Let D(t) be a continuous positive function[*] of t at most
$\pi/2$ and such that any arc of g for which

---
[*]We suppose that D(t), x(t) and y(t) have the period
2π in t.

$$t_0 \geq t \geq t_0 - D(t_0)$$

is simple. Any continuous branch of the many-valued
function

(18.2)   arc $\Big[z(t) - z[t - D(t)] \Big]$

will change by a quantity $2p\pi$ as t increases from 0 to 2π,
where p is an integer. It is supposed that g is described in the positive sense, as t increases.

It is clear that the above integer p is independent
of the difference $D(t)$ as conditioned above. Given two
choices, $D_0(t)$ and $D_1(t)$, one could introduce the deformation,

$$D(u, t) = uD_1(t) + (1 - u)D_0(t) \qquad (0 \leq u \leq 1),$$

where u increases from 0 to 1. We have

$$D(0, t) \equiv D_0(t), \qquad\qquad D(1, t) \equiv D_1(t).$$

For intermediate values of u, $D(u, t)$ lies between $D_0(t)$
and $D_1(t)$, and so is admissible if $D_0(t)$ and $D_1(t)$ are
admissible. If $D(t)$ in (18.2) is replaced by $D(u, t)$, the
angle in (18.2), properly chosen, becomes a continuous
function of t and u. The integer p must then be independent of u and thus independent of the choice of $D(t)$
among functions $D(t)$ conditioned as above. We term p
the angular order of g. If g is regular, $2\pi p$ is the angular variation of the tangent as g is traversed in the
positive sense.

A family of locally simply closed curves

$$z = f(t, a) = x(t, a) + iy(t, a)$$

(18.3)

$$(0 \leq t \leq 2\pi) \qquad (a_1 \leq a \leq a_2)$$

will be termed an <u>admissible</u> deformation, if $f(t, a)$ is
continuous in (t, a) and the curves of the family are
uniformly locally simple. Under this deformation the
curve $z = f(t, a_1)$ is admissibly deformed into the curve
$z = f(t, a_2)$. If e is a sufficiently small positive constant one can take $D(t) \equiv e$ for all the curves of the
family. The angle (18.2) will then vary continuously
with the parameter a. Thus the angular order p will remain constant throughout the deformation.

The angular order p of a positively sensed Jordan
curve is 1. For the same curve traced n times in the
positive sense, $p = n$; and, traced n times in the negative sense, $p = -n$. The figure eight traced in either
sense has the angular order 0. We shall presently see
that any two locally simple curves with the same angular
order can be admissibly deformed into each other.

The following lemma is needed.

<u>LEMMA</u> 18.1. <u>In a set S of uniformly locally simple
curves any curve which is sufficiently near a Jordan
curve g in the sense of Fréchet is simple.</u>

If the lemma were false there would exist a sub-
sequence g_n of curves of S tending to g in the sense of
Fréchet, with a multiple point P_n on g_n tending to a point
P of g. Suppose that every sub-arc of a curve of S of
diameter at most e is simple. Let h be an arc of g containing P in its interior, with a diameter less than $e/2$.
For all n sufficiently large there will be a sub-arc h_n
of g_n so close to h in the sense of Fréchet that the diameter of h_n is less than e, while the arc $g_n - h_n$ is
bounded from P. Such an arc h_n will be simple and for n
sufficiently large must contain P_n; otherwise P_n would

have a limit point other than P. But for n sufficiently
large, P_n on h_n cannot be a multiple point of g_n, since
for such n, h_n is simple and $g_n - h_n$ is bounded·from P.
This contradiction implies the lemma.

§19. Interior transformations

We shall consider transformations w = f(z) which
map $\bar{G}$ into the w-plane. These transformations shall be
interior (except for poles) in the neighborhood of each
point of G, and shall be continuous at points of (B).
The image of the boundary curve B_1 will be denoted by g_1.
We shall assume that the curves g_1 do not intersect w = 0.
Two sets of boundary conditions will be used.

I. Under Conditions I the transformation f(z) shall
be 1 - 1 in some neighborhood (relative to $\bar{G}$) of each
point of (B).

A point z_0 whose image w_0 is a branch point is
called a branch point antecedent. Such a point will be
called primary if f(z) has neither pole nor zero at z_0.
Branch point antecedents which are zeros or poles of
f(z) will be called secondary.

We shall study f(z) by studying the pseudo-harmonic
function

$$U(x, y) \equiv \log |f(z)|.$$

At each zero or pole of f(z), U has a logarithmic pole.
The function U is continuous at points of (B) since f(z)
does not vanish there. The following notation should be
recalled:

n(0) = The number of zeros of f(z) on G

n(∞)=   "   "   "   poles of f(z) on G

μ =   "   "   "   branch·point antecedents of

f(z) on G

M = The number of logarithmic poles of U on G

S = "   "   "   saddle points of U on G.

Poles, zeros, branch point antecedents and saddle points are counted with their multiplicities, but logarithmic poles are counted singly.

LEMMA 19.1. $n(0) + n(\infty) - \mu = M - S$.

The number M is the number of zeros and poles of $f(z)$ on G counted singly. The number S is the number of primary branch point antecedents counted according to their multiplicities. The number R of secondary branch point antecedents counted according to their multiplicities is the sum of the absolute orders of the zeros and poles of $f(z)$ diminished by 1 for each pole or zero, that is diminished in toto by M. Thus

$$R = n(0) + n(\infty) - M.$$

The total number of branch point antecedents is then

$$R + S = [n(0) + n(\infty)] - M + S.$$

On equating this number to μ, to which it is equal by definition of μ, one obtains the lemma.

The following lemma will be established in §28.

LEMMA 19.2. Corresponding to any locally simple, closed curve k there is a sequence k_n of regular closed curves which tend to k in the sense of Fréchet, which are uniformly locally simple, and which have the same angular order as k.

Theorem 19.1 is fundamental:

THEOREM 19.1. <u>Under Boundary Conditions</u> I

$$n(0) + n(\infty) - \mu = 2 - v + q - p,$$

<u>where</u> q <u>and</u> p <u>are respectively the sum of the orders with</u> <u>respect to</u> w = 0 <u>and the sum of the angular orders, of</u> <u>the images of the boundary curves under the interior</u> <u>transformation</u> w = f(z).

The formal structure of the proof may be seen by first supposing that the image curves g_i are regular. These curves lie in the (u, v) plane. In terms of the coordinates (u, v)

$$U(x, y) \equiv \log |w| \equiv \frac{1}{2} \log (u^2 + v^2) = V(u, v),$$

V being defined by this equation. Thus U satisfies the Generalized Boundary Conditions B. (See §14). That is, V(u, v) is locally of class C' and ordinary, and the boundary g_i is regular. Let J be the resultant vector index of V(u, v) along the boundary images g_i. It follows from Theorem 16.1 and Lemma 17.2 that

$$M - S = 2 - v + J = 2 - v + q - p.$$

It is immaterial whether the curves g_i have multiple points or not.

In general the boundary images will not be regular, but the boundaries B_i can be admissibly modified as follows so that this regularity condition holds.

We first simplify the problem by supposing that the curves B_i are circles. No generality is lost thereby, since a homeomorphic mapping of $\overline{G}$ onto a domain bounded by circles will not change the character of f(z) as an interior transformation nor the values of the integers

appearing in the theorem. The theorem will remain true
or false according as it is true or false on replacing
each boundary curve B_i by a concentric circle B^1 on G
sufficiently near B_i. To that end, B^1 need only be so
near B_i that

(a) the new boundaries B^1 do not intersect,

(b) the subregion G* of G bounded by the circles
B^1 contains all of the zeros, poles, and branch
point antecedents of $f(z)$,

(c) the transformation $f(z)$ is one-to-one on some
neighborhood (relative to $\overline{G}$)of each point of
$\overline{G}$ on or between B^1 and B_i.

When B^1 satisfies these conditions, a deformation of
B_i into B^1 through concentric circles will correspond
under $w = f(z)$ to an "admissible" deformation of the
image g_i of B_i into an image g^1 of B^1. The curve g^1 will
have the same order and angular order as g_i. Thus the
characteristic integers appearing in the theorem will be
unchanged. It accordingly suffices to prove the theorem
for the modified region G*.

By virtue of Lemma 19.2 there exists in the (u, v)
plane a regular closed curve g_i', arbitrarily near g^1 in
the sense of Fréchet, with the angular order of g^1, and a
norm e of local simplicity independent of the nearness of
g_i' to g^1. The antecedent B_i' of g_i' will have a norm of
local simplicity independent of its nearness to B^1 if B_i'
lies within the closure of a neighborhood of B^1 on which
$w = f(z)$ is locally 1 - 1. In accordance with Lemma 18.1,
B_i' must be simple if sufficiently near B^1. We can accordingly suppose B_i'.so near B^1 that the preceding three
conditions (a), (b) and (c) on B^1 are satisfied by B_i'
[except of course the condition that B_i' be a circle].
For the new region bounded by the curves B_i' the characteristic integers in the theorem will then be the same as
originally. But we have seen that the theorem holds for
regular boundary images. Hence the theorem holds in general.

§20.   First applications and extensions

Example 1. The fundamental relation

$$(20.1) \qquad n(0) + n(\infty) - \mu = 2 - v + q - p$$

under Boundary Conditions I is illustrated trivially by
the function z^m defined on $|z| \leq 1$ with $m > 1$. Here the
integers in (20.1) are respectively

$$m + 0 - (m - 1) = (2 - 1) + m - m.$$

Example 2. The most fruitful source of illustration
of (20.1) for $v = 1$ is found in simply connected sub-
regions R of a Riemann surface S spread over the w-plane,
with R bounded by a curve g which is simple on S. The
domain $\overline{R}$ can be mapped 1 - 1 and continuously onto the
disc $|z| \leq 1$. The resulting mapping function $w = f(z)$
is interior. If g passes through no branch point of S,
its projection g_1 on the w-plane is locally simple. Indeed, f is then locally 1 - 1 neighboring the boundary
$|z| = 1$.

As an example, consider the two-sheeted Riemann surface with branch points at $w = 0$ and $w = 2$. Let the
sheets S_1 and S_2, cut along the real axis from $w = 0$ to
$w = 2$, be joined in the usual way to make S. The point
at infinity has a one-sheeted neighborhood on S_1 and also
on S_2. The surface S is simply connected since both S_1
and S_2 are the homeomorphs of hemispheres with circular
boundaries corresponding to the cuts. Joining two such
hemispheres along their circular boundaries obviously
gives a sphere homeomorphic to S.

Any simple closed curve g on S divides S into two
regions each the homeomorph of a circular disc. We shall
take **g** as a curve whose projection on the w-plane is a

figure-eight with vertex at w = 1 and with loops which
encircle w = 0 and w = 2 respectively. (See Fig. 3,
where S_1 and S_2 are pictured separately.)

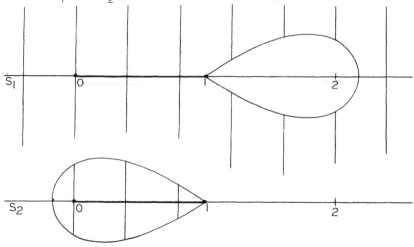

Figure 3.

    One has two choices of the region R bounded by g.
We have chosen the region shaded in Fig. 3. On sheet
S_1, R includes all of S_1 except the interior of the loop
on S_1. On S_2, R includes only the interior of the loop.
The region R is single-sheeted except over the loop encircling w = 0; in particular R is single-sheeted neigh-
boring the point at infinity. If w = f(z) maps $|z| \leqq 1$
onto $\overline{R}$ the integers in (20.1) are respectively

$$n(0) + n(\infty) - \mu = 1 + q - p$$
$$2 + 1 - 1 = 1 + 1 - 0.$$

    <u>Example</u> 3.   Divide S by a circle on S_1 of radius 3
with center at the origin. Let R be chosen as the region
which includes w = 0 on S_1. Then R includes all of S_2
and the substitution in (20.1) takes the form

$$2 + 1 - 2 = 1 + 1 - 1.$$

For any point $w = a$ not on the image (g) of (B), let $g(a)$ denote the total order with respect to a of (g). Relation (20.1) can then be extended as follows.

THEOREM 20.1.   Under Boundary Conditions I and for arbitrary points $w = a$ and $w = b$ not on the image (g) of (B),

(20.2) $n(a) + n(b) - \mu = 2 - v + q(a) + q(b) - p.$

In particular a may equal b; if one or both of the points a and b is infinite, (20.2) still holds, provided one sets $q(\infty) = 0$.

Since any finite point $w = a$ not on (g) can be taken as the origin, it appears from (20.1) that

(20.3) $n(a) + n(\infty) - \mu = 2 - v + q(a) - p.$

If $|a|$ is sufficiently large

$$n(a) = n(\infty) \qquad q(a) = 0.$$

Hence (20.3) implies the relation

(20.4) $2n(\infty) = \mu + (2 - v) - p.$

On multiplying the relation (20.3) by 2 and subtracting (20.4) one finds that

(20.5) $2n(a) = \mu + (2 - v) + 2q(a) - p$

for any a not on g. The relation (20.5) taken for b instead of a, and added to the relation (20.5) gives (20.2)
as stated, and the theorem is proved.

Relation (20.2) gives (20.5) on setting a = b, and
(20.4) on setting a = b = ∞. On subtracting relation
(20.5) for a, from (20.5) for b, one obtains the order
relation,

(20.6) $n(a) - n(b) = q(a) - q(b)$,

which specializes into

(20.7) $n(a) - n(\infty) = q(a)$.

Relation (20.2) implies each relation stated.

The order relation (20.6) has been established under
Boundary Conditions I. It can be immediately extended
as follows:

THEOREM 20.2. The order relation

(20.8) $n(a) - n(b) = q(a) - q(b)$

holds if Boundary Conditions I are replaced by the condition that f(z) be continuous on the boundary (B), and
w = a and w = b do not lie on (g). The boundaries can be
taken as Jordan curves.

To prove this theorem we can suppose that the boundaries (B) are circular and replace each boundary B_1 by a
nearby concentric circle B^1 on which there are no branch
point antecedents. This is always possible since the
branch point antecedents are isolated on G. We suppose
B^1 so near B_1 that the antecedents of w = a or b lie on
the modified region G*. Boundary Conditions I hold for

G*, and Theorem 20.2 implies that

(20.9) $n(a) - n(b) = q_1(a) - q_1(b)$,

where q_1 is the total order relative to the images g_1 of the modified boundary. But

(20.10) $q_1(a) = q(a)$ $q_1(b) = q(b)$

since the new circular boundaries B^1 can be deformed through concentric circles into the original boundaries B_1, and the image g^1 of B^1 will thereby be continuously deformed into the image g_1 of B_1 without intersecting $w = a$ or $w = b$. Relation (20.8) follows from (20.10) and (20.9).

The order relation is well known in the theory of meromorphic functions. In the special case in which there are no poles, relation (20.2) becomes $n(a) = q(a)$. For this relation in more general situations, see Kuratowski, p. 358. One can apply the order relation to generalize Rouché's theorem: (Titchmarsh, p. 116):

Rouché's Theorem. If $f(z)$ and $g(z)$ are analytic within and on a closed contour C and

(20.11) $|g(z)| < |f(z)|$

on C, then $f(z)$ and $f(z) + g(z)$ have the same number of zeros within C.

In the generalization, Theorem 20.3 below, the functions are merely continuous on the boundary, and interior within. The condition (20.11), which requires that $g(z)$ be nearer the origin $w = 0$ than $f(z)$, is unnecessarily restrictive. For example, z and iz have the same number of zeros within the circle $|z| = 1$ but do not satisfy (20.11). A generalization follows.

THEOREM 20.3. If for every real t for which $0 \leq t \leq 1$, $f(z) + tg(z)$ is an interior transformation of G which is continuous on (B), and if on (B)

$$(20.12) \qquad\qquad f(z) + tg(z) \neq 0,$$

then $n(0) - n(\infty)$ is the same for $f(z)$ as for $f(z) + g(z)$.

Condition (20.12) is satisfied when (20.11) is satisfied; for whenever (20.11) holds

$$|tg(z)| \leq |g(z)| < |f(z)|$$

so that (20.12) holds.

As t varies from 0 to 1, the boundary images under $w = f + tg$ vary continuously, never passing through $w = 0$. Hence the order q(0) of the boundary is independent of t. Thus

$$n(0) - n(\infty) = q(0)$$

both for $f(z)$ and $g(z) + f(z)$. Rouché's Theorem follows in case $n(\infty) = 0$ and $v = 1$.

The fundamental theorem of algebra is a trivial consequence of Rouché's theorem. See Titchmarsh, p. 118.

The following existence theorem will be useful.

THEOREM[*] 20.4. Corresponding to an arbitrary sensed, regular, analytic, closed curve (in general not simple) there exists a function $f(z)$ which is meromorphic on

---

[*]This theorem is included among the hitherto unpublished results of Morse and Heins.

$\overline{G} = [\,|z| \leq 1\,]$, <u>continuous</u> on B $= [\,|z| = 1\,]$ <u>and</u> <u>such</u> <u>that</u> the <u>image</u> <u>of</u> B <u>under</u> f <u>is</u> g <u>with</u> f' $\neq$ 0 <u>on</u> B.

Let g be referred to its arc length s. Without loss of generality we can suppose that the total length of g is 2π. The curve g will then have the form

$$w = u(s) + i\,v(s) = F(s) \qquad (F'(s) \neq 0)$$

where u(s) and v(s) are real for s real, and analytic, with a period 2π in the complex variable s in a neighborhood of the segment 0 $\leq$ s $\leq$ 2π of the real s axis. Subject to the transformation $\eta = e^{is}$, set

$$F(s) = H(\eta).$$

The function H(η) so obtained is real and analytic in η at each point η of the circle, |η| = 1, and maps this circle in a locally 1 - 1 way onto g. The function H(η) can be given a Laurent development,

$$H(\eta) = \sum_{-\infty}^{\infty} a_n \eta^n\,,$$

in an annulus which includes the circle |η| = 1 in its interior. Set

$$H_m(\eta) = \sum_{-m}^{m} a_n \eta^n.$$

If m exceeds a sufficiently large integer N, the antecedent in the η-plane of g in the w-plane under the transformation

$$w = H_m(\eta) \qquad (m > N)$$

is a simple, regular, analytic curve B_m. For m $>$ N
let $\eta_m(z)$ map the interior of B in the z-plane in a directly conformal manner onto the interior of B_m in the
η-plane. This map may be conformally continued without
singularity into a map which transforms B in a 1 - 1 '
manner onto B_m. For any m $>$ N the transformation

$$w = H_m[\eta_m(z)] \qquad \{|z| \leqq 1\}$$

maps B in the z-plane onto g in the w-plane, and satisfies
the theorem.

   In proving the preceding theorem no attempt has been
made to control the number of zeros and poles of the
transformation f of the theorem. Relations (20.4) and
(20.5) are of course necessary for any point a not on g.
The numbers n(a), n(∞), and μ are never negative, and
this imposes further conditions on the integers which
satisfy (20.4) and (20.5). The problem of determining
the extent to which one can prescribe both g and μ and
still satisfy the theorem is of considerable interest.

   A topological invariant. Let g be an arbitrary locally simple curve in the finite w-plane. Let the w-
sphere be subjected to a sense-preserving homeomorphism T
which carries g into a second curve g' in the finite
w-plane. Let a be an arbitrary point of the w-sphere not
on g, and let a' be the image of a under T. Let q(a) and
q'(a') be respectively the orders of a and a' with respect to g and g', and let p and p' be respectively the
angular orders of g and g'. We have the theorem.

   THEOREM 20.5. Under the homeomorphism T of the w-
sphere

(20.13)   2q(a) - p = 2q'(a') - p'

where a' is the image of a under T, and where the image g'
of g under T is assumed to be in the finite plane. The
orders (p, q) refer to g, the orders (p', q') to g'.

We shall presently show that every locally simple
curve g can be admissibly deformed (cf. §18), among
curves arbitrarily near g in the sense of Fréchet, into a
regular, analytic curve. During such a deformation the
orders appearing in (20.13) will be unchanged, provided
of course that the deformation has been on a neighborhood
of g which excludes a. It will accordingly be sufficient
to prove the theorem for regular, analytic curves g.

Corresponding to any such regular, analytic curve g,
Theorem 20.4 affirms the existence of a function f which
is meromorphic, on $\overline{G} = [|z| \leq 1]$, continuous on B =
$[|z| = 1]$ and such that the image of B under w = f(z) is
g. If T is of the form w' = F(w) then under T, f(z) may
be replaced by F[f(z)], again an interior transformation
of $\overline{G}$. If n', μ' refer to the new function Ff, then

(20.14) $n'(a') = n(a)$ $\mu' = \mu$.

Relation (20.5) for the new function takes the form

(20.15) $2n'(a') = \mu' + 1 + 2q'(a') - p'$.

Relation (20.15), taken with (20.14) and (20.5) gives
(20.13).

The integer 2q(a) - p is thus a topological invariant of the locally simple curve g under homeomorphisms T
of the sphere. It is not difficult to give a direct
proof of this fact independent of the theory of interior
transformations.

Relation (20.13) may be specialized in a number of
ways. In particular, let c be the antecedent of ∞ under T.

On setting a = c one sees that

$$q'(a') = q'(\infty) = 0,$$

and (20.13) yields the result

(20.16) $p = p' + 2q(c).$

Relation (20.16) combined with (20.13) yields the following corollary.

   COROLLARY. If T(c) = ∞ and a is an arbitrary point not on g, then

(20.17) $p = p' + 2q(c), \qquad q(a) = q'(a') + q(c).$

   Relation (20.17) is of the nature of a linear transformation from the pair [p, q] to the pair [p', q'] where the coefficients in the transformation depend only upon T. As a final specialization suppose that T transforms the finite plane into the finite plane. Then c = ∞ and q(c) = 0 so that (20.17) takes the form p = p', q = q'.
   An example. Let T be the transformation $w' = \frac{1}{w}$. Let g be the positively sensed unit circle $|z| = 1$. Then g' is the negatively sensed unit circle and

$$p = 1 \qquad p' = -1 \qquad c = 0 \qquad q(c) = 1.$$

If one sets a = 0, then a' = ∞ and q'(a') = 0, q(a) = 1. Relations (20.17) are satisfied.

CHAPTER IV

THE GENERAL ORDER THEOREM

§21. An example

The relation

(21.1) $2n(a) - \mu = 2 - v + 2q(a) - p$

has been established under the conditions that the interior transformation f be locally 1 - 1 neighboring[*] each
point of the boundary (B), and that the point a not lie on
the boundary images (g). In this section the hypothesis
that f be locally 1 - 1 neighboring each point of (B)
will be replaced by the hypothesis that the boundary
images be locally simple. With this change, μ in
(21.1) must be given a new interpretation. Theorem 24.2
and its proof were presented for the first time in lectures at Princeton University in December 1945.

The only relation of character similar to (21.1)
which the author has found is that of Stoilow (2).
Stoilow's is concerned with a region with a single boundary B. The image of B is assumed to be a closed Jordan
curve g in 1 - 1 correspondence with B by virtue of f.
The transformation f is assumed interior on $\overline{G}$. Apart
from notation, Stoilow affirms that

(21.1) $n(0) + n(\infty) - \mu = 1.$

---

[*]Relative to $\overline{G}$.

Stoilow seems to be assuming other conditions on f near
the boundary, as his proof shows. Without such addition-
al assumptions, the theorem would be false as we shall
indicate by an example. A first impression would be that
one had merely to let μ include the count of branch point
antecedents on B. This, however, would be incorrect if
one counted branch points in the ordinary way, as our
example will show. A more serious limitation, in the
case of a function which is meromorphic on G and merely
continuous on $\overline{G}$, is that the term "branch point" on B is
either without meaning or requires extensive preparatory
analysis.

As we shall see, the Stoilow hypothesis that g be
the 1 - 1 continuous image of B under f can be replaced
by the hypothesis that g be a locally simple image of B
for which

$$q(0) = p.$$

Relation (21.2) then holds if μ be properly interpreted.

The counter-example. Consider the transformation

$$(21.3) \qquad f(z) = \frac{1 + 3z^2}{z(3 + z^2)} \qquad \{|z| \leqq 1\}.$$

Within the unit circle f vanishes at just two points,

$$z = \pm \sqrt{\frac{1}{3}},$$

and has a pole at $z = 0$ only. Let $\overline{z}$ be the conjugate of
z. When $|z| = 1$, $z\,\overline{z} = 1$ and

$$|f| = |\overline{z}^2 f| = \left| \frac{\overline{z}^2 + 3}{z^2 + 3} \right| = 1$$

so that the image of the circle $|z| = 1$ is on the cir-
cle $|w| = 1$. It remains to show that the circle $|w| = 1$
is the 1 - 1 image of the circle $|z| = 1$. To that end
note that

$$f'(z) = \frac{3(1 - z^2)^2}{z^2(3 + z^2)^2} \, .$$

Thus f' has a double zero both at $z = 1$ and -1. But $z =$
1 and -1 are fixed points of f. Hence the open semi-
circles of $|z| = 1$ on which y is respectively positive
or negative are either the 1 - 1 images of themselves or
of each other. But the points $z = \pm$ i are fixed points
of f so that the mapping of the circle $|z| = 1$ into the
circle $|w| = 1$ is 1 - 1.

  The transformation f is thus interior on $\{|z| \leqq 1\}$
and transforms the circle $|z| = 1$ in a 1 - 1 manner onto
the circle $|w| = 1$. For points z on $|z| < 1$.

$$n(0) = 2, \qquad n(\infty) = 1, \qquad \mu = 0,$$

and relation (21.2) is not satisfied. For points z on
$|z| \leqq 1$, and with the ordinary interpretation of μ,

$$n(0) = 2, \qquad n(\infty) = 1, \qquad \mu = 4,$$

and (21.2) is again not satisfied.

§22. Locally simple boundary images

  Let f be an admissible transformation of $\overline{G}$, continuous on $\overline{G}$ and interior on G. Let B_i be a Jordan
curve bounding $\overline{G}$, given as the homeomorph of a circle
C with angular parameter t. Under Boundary Conditions
II the image g_i of B_i under f shall be the continuous

and locally 1 - 1 image of the circle C. It is not e-
nough that g_i "cover" a locally simple curve. Boundary
Conditions II require that the correspondence, estab-
lished by f between any arc of B_1 for which $|\Delta t|$ is suf-
ficiently small and its image arc on g_1, be 1 - 1.
Boundary Conditions II would not be satisfied if, as t
increased, the image point on g_1 remained fixed for any
interval of t, or if this image point traced an arc in
one sense and immediately thereafter retraced this arc
in opposite sense. For example g_1 might trace a circle
one or more times in several senses or even in one sense,
and still not satisfy Boundary Conditions II.

Under Boundary Conditions I (§19) each point P of B
is required to possess a neighborhood N relative to $\overline{G}$
whose w-image under f is in 1 - 1 correspondence with N;
more briefly we shall say that Boundary Conditions I require that f be locally 1 - 1 at each boundary point P.
Boundary Conditions I refer to a neighborhood of P relative to $\overline{G}$, Boundary Conditions II refer to a neighborhood
of P relative to (B). Both Boundary Conditions I and II
require that w = f(z) be 1 - 1 on the relative neighborhoods to which they refer. Boundary Conditions I imply
Boundary Conditions II, but the converse is not true.

We here introduce two definitions that will be of
use in the following section. Recall that a Jordan region R is a region bounded by a closed Jordan curve g.
The positive sense of g in the w-plane is the sense corresponding to which the order[*] of each interior point is 1.
A simple sensed arc b on the Jordan curve g will be said
to have the Jordan region R on its left or right according as the sense of b agrees or disagrees with the positive sense of g.

Let W be an (m + 1)-leaved branch element each sheet
of which covers a Jordan neighborhood N of a point w_0 of
the w-plane. Let b be a simple, sensed arc through w_0
which separates N into two Jordan regions of which N_1

[*] That is the "order of g relative to the interior point".

lies to the right of b. The connected subregion of W obtained by removing a single sheeted copy of N_1 from W
will be called a partial branch element of multiplicity m.
Zero is admitted as a value of m. A semi-circle with w_0
on the center of its diameter would be a partial branch
element of multiplicity c. The continuous image under
the transformation

$$(22.1) \qquad\qquad w - w_0 = re^{\theta i}$$

of the pairs (r, θ) for which

$$(22.2) \qquad 0 \leq \theta \leq (2m + 1)\pi, \qquad 0 \leq r \leq r_0, \qquad (r_0 > 0)$$

is representable as a partial branch element E_m of multiplicity m. It is clear that any two partial branch elements with branch point w_0 and multiplicity m admit a
homeomorphism in which w_0 corresponds to itself and
points which cover the same point w correspond to points
of like character. One could in fact define a partial
branch element of multiplicity m as the continuous image
of the preceding element E_m under a homeomorphism which
carries distinct sheets into distinct sheets and points
which cover the same point w into points of like character.
      The multiplicities of partial branch elements can be
formally determined in certain cases. Suppose that z_0 is
on (B) and that f(z) is analytic at z_0, while $f(z)-f(z_0)$
vanishes at z_0 to the nth order. If two regular arcs of
(B) intersect at z_0 so as to make an angle A at z_0 within
G, and if

$$2m\pi < nA < 2(m + 1)\pi,$$

then m is the multiplicity of the partial branch element

at $f(z_0)$. If the boundary is regular at z_0 and $n =$
$2r + 1$, then $m = r$. In general, the multiplicity m cannot
be determined by such formal processes, but is <u>a priori</u>
characteristic of the interior transformation.

An <u>example</u>. In the example of the preceding section, the points $z_0 = \pm 1$ correspond to partial branch
elements of multiplicity 1. The difference $f(z) - f(z_0)$
vanishes to the third order for $z_0 = \pm 1$.

§23.   <u>The existence of partial branch elements</u>

We shall show that the hypothesis that the boundary
images g_i are locally simple implies that the transformation is "locally 1 - 1" (cf. §22) at each point of (B)
with the possible exception of a finite number of points
P of (B), and that the image under f of some neighborhood,
relative to $\overline{G}$, of each such exceptional point P is a
<u>partial branch element</u>. The truth or falsity of such a
theorem will clearly be independent of any sense-preserving homeomorphism F of the w-sphere, leaving the point ∞
fixed, with which f may be composed to form a new interior
transformation $F[f(z)]$. It is an essential advantage of
the topological methods of proof which we follow that no
generality is lost upon replacing f by the new interior
transformation.

The following lemma affords a particular homeomorphism F.

LEMMA 23.1. <u>Corresponding to any simple arc</u> k
<u>(closed) there exists a sense-preserving homeomorphism</u> F
<u>of the</u> w-<u>sphere, which leaves the point</u> ∞ <u>fixed and maps</u>
k <u>into a straight arc.</u>

To obtain F, one joins the end points of k by a
simple arc so as to form a Jordan curve g bounding a
Jordan region R. Let $\overline{R}$ be mapped by a homeomorphism[*]H
onto a rectangle Q in such a manner that k goes into a

---
[*]Sense-preserving

side k_2 of Q. Let the complement CR of R on the w-sphere be mapped by a homeomorphism H' onto $\overline{CQ}$ in such a manner that the maps H and H' are identical on g. (Cf. Lemma 6.1.) We suppose moreover that H' leaves the point ∞ fixed. The maps H and H' combine to define the homeomorphism F affirmed to exist in the lemma.

In this section polar coordinates (r, θ) will be used in the w-plane with pole at w = 0. We are assuming that the boundary images g_1 under w = f(z) are locally simple. Attention will be focused on an arc h (closed) of (B) so small in diameter that its image under f is a simple arc h^f in 1 - 1 correspondence with h. Use will be made of the preceding lemma by virtue of which h^f may be assumed to be <u>straight</u>.

<u>The</u> <u>analysis</u> <u>will</u> <u>be</u> <u>made</u> <u>simpler</u> <u>by</u> <u>supposing</u> <u>that</u> h^f <u>lies</u> <u>on</u> <u>a</u> <u>ray</u> θ = $θ_0$ <u>and</u> <u>does</u> <u>not</u> <u>intersect</u> w = 0.

We begin with the following lemma.

LEMMA 23.2. <u>The</u> <u>antecedents</u> <u>on</u> G <u>of</u> <u>branch</u> <u>points</u> <u>of</u> f^{-1} <u>have</u> <u>no</u> <u>limit</u> <u>point</u> <u>at</u> <u>inner</u> <u>points</u> <u>of</u> <u>the</u> <u>bound-</u> <u>ary</u> <u>arc</u> h, <u>and</u> <u>the</u> <u>points</u> <u>of</u> h <u>at</u> <u>which</u> f <u>fails</u> <u>to</u> <u>be</u> <u>locally</u> 1 - 1 <u>are</u> <u>isolated</u>.

The image h^f of h under f is on the ray θ = $θ_0$ so that the pseudo-harmonic function

$$U(x, y) = \log |f(z)| = \log r$$

will have no extremum at an inner point of h. In accordance with the proof of Theorem 7.1 the saddle points of U on $\overline{G}$ cannot cluster at any inner point of h. Hence the antecedents on G of branch points of f^{-1} cluster at no inner point of h. It remains to show that the points of h at which f fails to be locally 1 - 1 are isolated.

Let P be any inner point of h at which just one level curve X of U terminates. (See Fig. 4.) The points of h which are not of this type are isolated on h (cf. §7). We shall show that f is locally 1 - 1 at P.

Set $U(P) = c$. Let X be so limited that each of its
points are ordinary points of U. Let Y be a U-trajectory
through the end point of X not on h, with U taking on the
values

$$c - e \leq U \leq c + e \qquad\qquad (e > 0)$$

on Y. If e is sufficiently small there will be one and
only one level curve (r constant thereon) passing from an

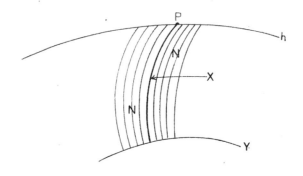

Figure 4.

arbitrary point on Y to a point on h, as seen in §7, and
these level curves will cover a neighborhood N of P relative to $\overline{G}$, with one and only one curve through each point
of $\overline{N}$. On each such level curve the transformation f is
1 - 1. For there can be no antecedents of (ordinary)
branch points on $\overline{N}$ if there is only one level curve
through each point of $\overline{N}$.

The map f is 1 - 1 for z on $\overline{N}$. In fact, each level
arc on $\overline{N}$ corresponds in the w-plane to an arc of a circle
on which r is constant, while arcs on $\overline{N}$ emanating from
different points of Y are at different levels and so correspond to arcs of circles with different radii r. As
noted previously, the correspondence between a particular
level arc on $\overline{N}$ and its image is 1 - 1. Thus f is 1 - 1 on
$\overline{N}$. The lemma follows.

The inverse of $w = f(z)$ for z on $\overline{N}$ is continuous.
This follows from a well-known theorem which states
that a 1 - 1 continuous transformation of a bounded closed
plane set has a continuous inverse. See v. Kerékjártó,
p. 34.

Inner points P of h are either ordinary with respect
to U on $\overline{G}$, or else are saddle points with respect to U on
$\overline{G}$. In the former case we have just seen that f is
locally 1 - 1 at P. The following lemma completes the
analysis in the small.

LEMMA 23.3.· Each saddle point of U on $\overline{G}$ at an inner point of h possesses a neighborhood (relative to $\overline{G}$)
whose Riemann image is a partial branch element.·

We consider the case in which P is a saddle point of
U. Let U = c at P, and let S (open) be a sector of a canonical neighborhood of P relative to U. The argument
used in the proof of Lemma 23.2 suffices to show that
there exists a neighborhood V of P, relative to $\overline{S}$, whose
points, not at the level c, are covered by a family of
level arcs at different levels of U, with just one arc
through each point of S on V. Let V^f then be the image
of V under f. We distinguish between the cases in which
S is and is not of boundary type (cf. §7.) Set $(r, \theta) =$
(r_0, θ_0) at f(P).

Case A. The sector S is not of boundary type.
Suppose that S is below c, or in terms of r, below r_0.
In Case A there are two arcs, a and b, on the boundary of
S at the level r_0 with end points at P. As z approaches
P on a and recedes from P on b, f(z) traces an arc of the
circle $|w| = r_0$ approaching f(P) on $|w| = r_0$ from one
side and receding from f(P) on the other. That f(z) does
not approach and leave f(P) on the same sub-arc of $|w| =$
r_0 may be seen as follows: Such a double tracing of an
arc of the circle $w = r_0$ occurs only if a continuous
branch of

$$\theta(z) = \text{arc } f(z),$$

evaluated as a function of z on the arc $(a\ b)$, has an extremum at P. In such a case the level arcs on V on
which r is sufficiently near r_0 but below r_0 would bear
a similar extremum of θ; that is, the correspondence between these level curves on V below r_0 and their images
could not be 1 - 1 no matter how near the level r was to
r_0. This is contrary to the fact that there are no critical points of U on S, and that f. is accordingly locally
1 - 1 on S.

Thus the property of the level curves of V below r_0
of being in 1 - 1 correspondence with their images ex-
tends to the curve at the level r_0 on the boundary of V.
It follows that there exists a neighborhood V of P relative to $\overline{S}$ whose image V^f under f is of the form,

(23.1)
$$r_0 \gtreqless r > r_0 - e$$

$$\theta_0 + e > \theta > \theta_0 - e,$$

with $\overline{V}$ and $\overline{V}^f$ homeomorphic under f, provided e is a sufficiently small positive constant.

On the other hand, S may be <u>above</u> r_0. Then V^f may
be taken in the form,

(23.2)
$$r_0 + e > r \gtreqless r_0$$

$$\theta_0 + e > \theta > \theta_0 - e,$$

with $\overline{V}$ and $\overline{V}^f$ homeomorphic under f.

Case B. <u>The sector S is of boundary type</u>. Recall
that the image of h under f is an arc h^f on the ray $\theta = \theta_0$. Let h^f be given a positive sense derived under f
from the positive sense of h. Let k_1 and k_2 be the two

arcs of h^f into which h^f is divided by $f(P)$, with k_1 preceding k_2 in the positive sense of h^f. Case B breaks up into four sub-cases as follows.

I.   S is above r_0, k_1 is on $\bar{s}^f$.

II. S is below r_0, k_1 is on $\bar{s}^f$.

III. S is above r_0, k_2 is on $\bar{s}^f$.

IV. S is below r_0, k_2 is on $\bar{s}^f$.

Analysis similar to that presented under Case A yields the following result. For a sufficiently small positive constant e the image V^f of a neighborhood V of P relative to $\bar{S}$ may be taken in the respective forms,

I. $r_0 + e > r \geqq r_0$

$\theta_0 \geqq \theta > \theta_0 - e$

II. $r_0 \geqq r > r_0 - e$

$\theta_0 + e > \theta \geqq \theta_0$

III. $r_0 + e > r \geqq r_0$

$\theta_0 + e > \theta \geqq \theta_0$

IV. $r_0 \geqq r > r_0 - e$

$\theta_0 \geqq \theta > \theta_0 - e$

It is observed that $\theta_0 \geqq \theta$ in I and IV while $\theta_0 \leqq \theta$ in II and III. An explanation of this in the typical Case I is as follows: On the boundary of V^f in I, k_1 is followed by an arc k' on which $r = r_0$. Taken in its positive sense k_1 has V^f on its left (cf. §22), since its

antecedent on h has V on its left, and f preserves orientation. But k' follows k_1 on the boundary of V^f and will
accordingly have V^f on its left. Since $r \geq r_0$ on V^f in I
k' will have V^f on its left only if θ decreases from θ_0
on k'. Hence in the representation I of V^f, $\theta_0 \geq \theta$ must
hold rather than $\theta_0 \leq \theta$.

The preceding neighborhoods V of P relative to the
respective closed sectors $\overline{S}$ of a canonical neighborhood
of P can be successively joined in clockwise order to
form a neighborhood N of P relative to $\overline{G}$. The respective
images V^f of these neighborhoods V joined in corresponding clockwise order around f(P) yield the image N^f of N
under f. In this union each V^f except the last will be
joined to its successor along an arc of the circle $|w| = $
r_0 with an end point at f(P). Successive images V^f will
lie on opposite sides of the circle $|w| = r_0$, and the
initial and final images V^f will lie on opposite sides of
this circle, since h^f is a straight arc passing through f(P)
on θ = θ_0. The angle at f(P) on an image V^f under Case
A is π, and under Case B is π/2. The total angle at f(P)
on N^f is of the form (2m + 1)π, and V^f may be regarded as
a partial branch element of multiplicity m.

This completes the proof of the lemma.

§24.   The order, angular order theorem

The lemmas of the last section apply to any arc h of
(B) so small in diameter that its image under f is a
simple arc h^f in 1 - 1 correspondence with h. A finite
set of such arcs suffices to cover the whole of (B) in
such a manner that every point of (B) is an inner point
of such an arc. We draw the following conclusion.

THEOREM 24.1.   If the images (g) under the interior
transformation f of the boundaries (B) are locally simple then f is locally 1 - 1 relative to $\overline{G}$ at all but a

finite number of points on the boundary (B), and each exceptional point on (B) has a neighborhood relative to $\overline{G}$
whose Riemann image under f is a partial branch element.

If F(w) is a sense-preserving homeomorphism of the
w-sphere, with the point w = ∞ a fixed point, then the
function $f_1(z) \equiv F[f(z)]$ will be said to be w-<u>equivalent</u>
to f. If f is defined on $\overline{G}$ and has locally simple boundary images, f_1 is defined on $\overline{G}$ and has locally simple
boundary images. The total angular order p of these .
images is the same for f_1 as for f, as has been seen in
(20.18). We shall use this result in proving the following principal theorem.

THEOREM 24.2. <u>If the boundary images</u> (g) <u>are locally simple and do not intersect the point</u> w = a,<u>then</u>

$$(24.0) \qquad 2n(a) = 2 - v + \mu + 2q(a) - p \, ,$$

where n(a), q(a), v, <u>and</u> p <u>are as previously defined, and</u>
μ <u>is the sum of the multiplicities of the branch elements
of</u> f^{-1}, <u>evaluated as in</u> §22 <u>in the case of partial branch
elements</u>.

Let P_k (k = 1, ..., n) be the antecedents on (B) of
boundary branch points whose multiplicites m_k are positive. The transformation f is locally 1 - 1 relative to
$\overline{G}$ at each other point of (B). Let μ' be the sum of the
multiplicities of the ordinary branch points with antecedents on G. We shall replace each curve B_i of (B) by a
Jordan curve which traces B_i in its positive sense making
a detour h_k on $\overline{G}$ around each point P_k on B_1.
The detour h_k will be a simple arc on $\overline{G}$ which replaces a simple arc of (B) through P_k, and will be more explicitly defined as follows. Given P_k let $f_k(z)$ be an interior transformation which is w-equivalent to f and which

is such that the image under f_k of some arbitrarily small
simple boundary arc of (B) through P is a straight arc b_k
in the w-plane. That f_k exists follows from Lemma 23.1.
There exists a neighborhood N_k of P_k relative to $\overline{G}$ such
that the Riemann image of N_k under f_k is a partial branch
element of multiplicity m_k. Set $w_k = f_k(P_k)$. Let the detour h_k around P_k on $\overline{G}$ be taken so that its image under
f_k is circular in form, with center at w_k, and subtends
an angle at w_k of $-(2m_k + 1)\pi$, beginning at a point of b_k
which precedes w_k on b_k, and ending at a point of b_k
which follows w_k on b_k.

The respective detours h_k can be made so near the
points P_k that for the region G* which is enclosed by the
new boundaries the numbers n(a), q(a), and μ' for f will
be the same as for G. Boundary Conditions I will be satisfied by f and the new boundaries. Let p' be the total
angular order of the images of the new boundaries, under
f. Since Boundary Conditions I are satisfied by f on $\overline{G}*$

(24.1) $2n(a) = 2 - v + \mu' + 2q(a) - p'.$

We shall show that

(24.2) $p = p' + \mu - \mu'.$

This amounts to showing that

(24.3) $p = p' + \sum m_k.$

If the boundaries (B) are altered by taking the detour h_k it follows from the nature of the image of h_k under f_k that the angular order of the boundaries will
thereby be decreased by m_k; from the invariant character
of angular order under our w-homeomorphisms, as indicated
by (20.17), it is clear that the angular order of the

boundary images under f will be decreased by the same
amount. The boundaries (B) can be altered by taking the
detours h_k in succession. On summing the changes in
angular order (24.3) results. Hence (24.2) holds and
(24.0) results on adding (24.1) and (24.2).

The proof of the theorem is complete.

Example. The function $f(z)$ given in (21.3) has two
boundary branch points with antecedents $z = \pm 1$. The corresponding partial branch elements have multiplicities 1,
and the relation,

$$n(0) + n(\infty) - \mu = 1 \; ,$$

is satisfied in the form,

$$2 + 1 - 2 = 1.$$

§25.   Radó's theorem generalized

A theorem in Titchmarsh

The theorem of Radó states that there exists no
$(1, m)$ directly conformal map $w = f(z)$ of G onto itself
when $m > 1$ and the number of boundaries $v > 1$. Branch
points are not excluded. In such a map each point $w = a$
on G is such that $n(a) = m$. It is relatively easy to
show, as Radó does, that the map $f(z)$ can be extended so
as to be continuous on (B), and such that each curve B_i is
mapped onto some curve B_j of (B) covered m times. The
angular order of the exterior boundary of G covered m
times is m. The other boundaries are covered m times but
in the opposite sense relative to their interiors; hence
their angular orders as boundary images (g) are all $-m$.
The total angular order of (g) is thus $(2 - v)m$. Our
generalization of Radó's theorem is confined to the case

where $v > 2$, and is as follows:

THEOREM 25.1. In the case in which the number of
boundaries exceeds 2, $n(\infty) = 0$, and the boundary images
(g) are locally simple, it is impossible that the total
angular order of the boundary images be $(2 - v)m$ with
$m > 1$ for any interior transformation defined on $\overline{G}$.

No assumption is made here as to the values of $n(a)$.
The boundary images may intersect themselves and each
other. In the case where (B) is a set of circles and f
is analytic, it is not assumed that $f(z)$ is analytic on
(B).

We make use of Theorem 24.2 according to which

$$2n(a) = \mu + 2 - v + 2q(a) - p$$

for a not on (g). We set $a = \infty$ and $p = (2 - v)m$. It
follows that

(25.1) $\mu = (v - 2)(1 - m)$.

Since $\mu \gtrless 0$ this condition is impossible when $v > 2$ and
$m > 1$. The theorem follows.

When $v = 2$ it is seen from (25.1) that $\mu = 0$. For
$v = 2$ and $m > 1$, a $(1, m)$ interior transformation of an
annulus onto itself is clearly possible. Here is a difference between the theory of interior transformations and
meromorphic functions, because as Radó has shown there is
no directly conformal $(1, m)$ transformation of the annulus
onto itself for $m > 1$. Radó establishes this by extending
the transformation over the whole plane by repeated re-
flections in the circular boundaries, arriving at the result that the only $(1, m)$ directly conformal transformation of the annulus into itself is a rotation.

It appears that when v $>$ 2 Radó's theorem is essentially topological, but that when v = 2 it belongs properly to conformal mapping theory.

A theorem in Titchmarsh (p. 122) is of interest. A
curve on which $|f(z)|$ equals a positive constant is there
called a level curve of f.

THEOREM 25.2. If f is analytic within and on a regular Jordan level curve C and has n zeros within C, then
f' has n - 1 zeros within C and never vanishes on C.

The proof can be given with the aid of Theorem 24.2,
but an even simpler proof is obtained on setting

(25.2)   U(x, y) = log $|f(z)|$

and making use of Theorem 13.1. The harmonic function U
assumes a constant maximum on C since U becomes negatively infinite at each of the zeros of f. In the relation

$$M - S = 1 + I$$

M = n and I = 0 in accordance with Theorem 13.1. Hence
S = n - 1. Thus f' has n - 1 zeros within C.

U can have no critical point on its level curve C
without assuming its maximum on other level curves crossing C. This is impossible. Hence f' never vanishes on C.

The following generalization of Theorem 25.2 goes
deeper in that f is taken as interior on R instead of
analytic. The strong assumption that f is analytic on C
is replaced by the condition that the image of C under f
is locally simple. The conclusion that f' does not vanish on C is replaced by the conclusion that f is locally
1 - 1 relative to $\overline{R}$ at each point of C. The theorem follows.

THEOREM 25.3. Let C be a Jordan curve bounding a
finite region R. If f(z) is an interior transformation
of R and is defined,finite and continuous on $\overline{R}$, if C is a
level curve of f and the image of C under f is locally
simple, if finally, f has n zeros on R, then μ = n - 1
and f is locally 1 - 1 relative to $\overline{R}$ at each point of C.

One makes use of the pseudo-harmonic function U of
(25.2) and concludes exactly as in the proof of Theorem
25.2 that f has n - 1 branch point antecedents within C.
To show that f is locally 1 - 1 at each point of C
use is made of Theorem 24.2 by virtue of which

$$(25.3) \qquad 2n(0) = 1 + \mu + 2q(0) - p.$$

Here n(0) = n = q(0). The image of C is a circle $|w|$ =
r_0 traced n times without reversal of sense,since this
image is locally simple. Thus p = n, μ = n - 1. Hence
there are no partial branch elements of positive multiplicity and the transformation f is locally 1 - 1 at each
point of C.

CHAPTER V

DEFORMATIONS OF LOCALLY SIMPLE CURVES
AND OF INTERIOR TRANSFORMATIONS

§26. Objectives

It has been seen that the order q (with respect to
w = 0) and angular order p of a locally simple closed
curve g are significant characteristics of g when g appears as the image of a boundary of G under an interior
transformation f. There exist locally simple curves for
which p and q are arbitrary integers as we shall see.
The meaning of p will be revealed by the theorem that any
locally simple sensed curve with the angular order p can
be admissibly deformed (cf. §18) into a figure eight if
p = 0, and if p ≠ 0 into a circle C traced p times, taking account of the positive sense of C. No two of these
models can be admissibly deformed into each other.

The meaning of the pair (p, q) will be brought out
by the use of 0-deformations of g. 0-deformations of g
are admissible deformations of g in which neither g nor
its images intersect the origin 0. Corresponding to any
pair of integers (p, q) there is a canonical curve g(p,q)
into which a curve with the invariants (p, q) can be 0-
deformed, and no two canonical curves g(p, q) corresponding to different pairs (p, q) can be 0-deformed into each
other.

The meaning of the invariant p of a locally simple
closed curve is made clearer by the definition of the

<u>product</u> of the deformation classes of such curves. The
groups G of such classes with respect to multiplication
is shown to be isomorphic with the additive groups J of
integers, with a deformation class c in G corresponding to
the integer p in J which equals the angular order of the
members of c. The unit class in G is composed of locally
simple curves admissibly deformable into the figure
eight. The groups G of 0-deformation classes of locally
simple curves which do not intersect w = 0,is similarly
shown to be isomorphic with the additive groups of pairs
of integers (p, q). The unit class in G turns out to be
the class of locally simple curves which are 0-deformable
into a figure eight neither loop of which encircles the
origin w = 0.

   The study of the deformation classes of locally
simple curves is a necessary introduction to the study of
deformation classes of interior transformations or meromorphic functions which have a prescribed set of zeros,
poles, and branch points, on a fixed region G. In §34
we shall refer to this problem and to the results already
obtained. It is sufficient to say that it leads to the
heart of the modern theory of meromorphic functions,in-
volving the "normal families" of Montel and the extensions
of the Picard Theorem. See Morse and Heins (2). Apart
from this, the substance of this chapter is a self con-
tained development of the "homotopy" theory of locally
simple closed curves.

   The following section introduces the concept of
μ-<u>length</u> of curves. This is a necessary aid in the representation of deformations of closed curves. By means of
the theorems there proved the possibility of an admissible
representation of such deformations is made to depend upon
certain conditions easily recognized and in general fulfilled, thus avoiding a detailed description of the representation in each case at hand.

§27.   The μ-length of curves.

Parameterization of curves by means of arc length
fails when the length is infinite. Even when finite, the
length need not vary continuously with the arc. To meet
this need a special parameter called μ-length (see Morse
(2)) is introduced. It is an extension of a function of
sets defined by Whitney (1). Whitney's definition applies to families of non-intersecting curves. The present extension is not so restricted and in this difference
lies the most of the difficulty. See also Fréchet (2).

The curves considered lie on a metric space M with
points p, q, r, etc. Distance between points p and q
will be denoted by pq. Here pq = 0 if and only if p = q,
while

$$(27.1) \qquad\qquad pq \leqq pr + qr.$$

That pq = qp follows readily from (27.1).

Let t be a number on a closed interval

$$(0 \leqq t \leqq a) \qquad\qquad (0 < a).$$

Let f(p, t) be a single-valued numerical function of p
and t, for p on M and t on (0, a). The function f(p, t)
is termed $\underline{continuous}$ at (p_0, t_0) if f(p, t) tends to
$f(p_0, t_0)$ as a limit, as p tends[*] to p_0 on M and t tends
to t_0. The continuity of a point function $\phi(p, t) \subset M$
is similarly defined.

A continuous map p(t) of t on (0, a) into a point
p(t) of M is termed a curve, or more precisely a parameterized curve (p-curve). We shall admit only those p-
curves for which p(t) is $\underline{constant}$ $\underline{on}$ $\underline{no}$ $\underline{sub\text{-}interval}$ of
(0, a). A second such p-curve

$$p = q(u) \qquad\qquad (0 \leqq u \leqq b)$$

---

[*]More precisely as (p, t) tends to (p_0, t_0).

will be said to define <u>the same curve</u> on M as p = p(t)
if there exists a sense-preserving homeomorphism between
the intervals (0, a) and (0, b) by virtue of which q(u) =
p(t).

   <u>Definition of</u> μ-<u>length of a curve</u>. Let a curve h.
have a representation p(t). Let T

(27.2) $t_1 \leqq t_2 \leqq \cdots \leqq t_n$ (n > 1)

be a set of values of t on the interval (0, a),and let

(27.3) $(p_1, p_2, \ldots, p_n)$

be the corresponding <u>admissible</u> set of points p(t) on M.
We consider the number

$$\min_i p_i\, p_{i+1} \quad (i = 1, \ldots, n\text{-}1).$$

As the t-values (27.2) vary on (0, a) we introduce the
number

$$m_n(h) = \max_T\,[\min_i p_i\, p_{i+1}].$$

Following Whitney set

$$\mu_h = \frac{m_2(h)}{2} + \frac{m_3(h)}{4} + \frac{m_4(h)}{8} + \cdots.$$

We term μ_h the (total) μ-length of the curve h. Observe
that the sum of the numerical coefficients is 1.

   Certain properties of $m_n(h)$ and μ_h will be enumer-
ated. Let d be the diameter of h. Then

(a) $m_n(h) \leq d$ and $\mu_h \leq d$

(b) $m_n(h)$ tends to 0 as n becomes infinite

(c) $m_n(h) \geq m_{n+1}(h)$.

Statements (a) and (b) are obvious. Statement (c) is proved as follows. A sequence p_i of (n + 1) vertices on h becomes a sequence of n vertices on removing one vertex p_k. Since n > 1 the vertex removed can be chosen so as not to affect the value of

(27.4) $\min_i p_i\, p_{i+1}$ (i = 1, ..., n).

It follows that the values of (27.4) to be maximized by varying the n values t_i will include all the values to be maximized by varying n + 1 values t_i, so that (c) will hold.

Let $\mu(t)$ be the μ-length of the p-curve on h defined by parameter values on the interval (0, t), with $\mu(0) = 0$.

(d) The μ-length $\mu(t)$ is a continuous function of t.

Corresponding to a sequence of n values t_i' on the interval (0, t') with t' > 0 one can choose a sequence of n values t_i'' so as to divide an interval (0, t") near (0, t') in ratios equal to those in which (0, t') is divided by the values t_i'. Every admissible sequence of n values t_i'' of t on (0, t") will be obtained in this way, and can be regarded as determined by the corresponding sequence of values t_i'. It is thus clear that

$$m_n(t) = \max_T\; [\min_i p_i\, p_{i+1}]\ (i = 1,..\ ,n\text{-}1)$$

will vary continuously with t. The continuity of $\mu(t)$ follows from the uniform convergence of the series defining $\mu(t)$.

(e)   The μ-length is an increasing function of t.

It is immediately clear that $\mu_n(t)$ never decreases as as t increases, since each sequence of n values $t_i^!$ on (0, t') is also on (0, t") for t" $>$ t'. To show that $\mu(t")$ $>$ $\mu(t')$ we shall first show that for n sufficiently large,

(27.5) $m_{n+1}(h") \geqq m_n(h')$,

where h' and h" refer respectively to the p-curves p(t) with intervals (0, t') and (0, t") respectively. Without loss of generality we can suppose that

(27.6) $p(t') \neq p(t")$

since a slight decrease of t" will insure this, and since relation (27.5) for the altered t" insures its truth for the original and larger t". With (27.6) holding, it follows from the triangle inequality,

(27.7) $0 < p(t') \, p(t") \leqq p(t) \, p(t') + p(t) \, p(t")$,

that the right-hand member of (27.7), considered as a function of t on the interval ($0 \leqq t \leqq t'$), has a positive minimum 2c.

Corresponding to an admissible set of n values $t_i^!$ of t on (0, t'), an admissible set of n + 1 values $t_i^"$ on (0, t") will be obtained by adding $t_{n+1}^"$ as t' if

$$p_n \, p(t') \geqq c,$$

and as t" otherwise. In the latter case

$$p_n \, p(t") \geqq c$$

by virtue of the definition of c. In either case

$$p_n\ p_{n+1} \gtrless c.$$

If n is sufficiently large, say $n \geq N$, then on h'

$$\max_i \min p_i\ p_{i+1} < c \qquad (i = 1, \ldots, n-1)$$

so that the addition of a new point p_{n+1} on h" will certainly not lead to a smaller max min $(i = 1, \ldots, n > N)$. Hence (27.5) holds as stated.

However, for some n sufficiently large,

$$(27.7.1) \qquad m_{n+1}(h') < m_n(h'),$$

since $m_n(h')$ tends to zero with $1/n$. A comparison of (27.5) and (27.7.1) shows that for some sufficiently large n

$$(27.8) \qquad m_{n+1}(h') < m_{n+1}(h'').$$

Since (27.8) with the equality added holds for all n it follows that

$$\mu(t') < \mu(t''),$$

and the proof of (e) is complete.

By virtue of (d) and (e) the relation between an admissible parameter t and the μ-length $\mu(t)$ is 1 - 1 and continuous. Hence the given curve h admits a representation

$$(27.9) \qquad p = q(\mu) = p[t(\mu)] \qquad [0 \leq \mu \leq \mu(a)]$$

where $t(\mu)$ is the inverse of $\mu(t)$. To indicate the de-
pendence of this representation on the curve h as well as
on the μ-length the representation will be written more
explicitly in the form

(27.10) $p = q(h, \mu)$ $[0 \leq \mu \leq \mu_h]$.

It is at once clear that the function $q(h, \mu)$ and end
value μ_h are independent of the particular p-curve used
to represent h.

A curve h is a class of p-curves H whose representations are deducible one from the other under homeomorphisms of their parameter. The Fréchet distance HK between
two p-curves is well defined, and by virtue of its form
is independent of admissible changes of parameter in H
and K. We can accordingly set hk = HK and be assured
that this definition of the distance between two curves h
and k is independent of admissible representations H and
K of h and k respectively. The distance hk satisfies the
relation

$$hk \leq hr + kr.$$

Moreover hk = 0 if h = k.

A pair (h, μ) in which μ is on the interval $(0, \mu_h)$,
and h on the space F of curves with a Fréchet metric, will
be termed admissible. We shall regard $q(h, \mu)$ as a function of admissible pairs (h, μ). A first theorem of im-
portance follows:

THEOREM 27.1. If h and k are curves for which hk $<$
e, then

(27.11) $|\mu_h - \mu_k| \leq 2e.$ $(e > 0)$.

Corresponding to a sequence of n points p_i on h, there exists an admissible sequence of n points r_i on k such that

$$p_i \; r_i < e \qquad (i = 1, \; 2, \; \ldots, \; n),$$

and each admissible sequence r_i on k can be so obtained. It follows from the triangle inequality that

$$|p_i \; p_{i+1} - r_i \; r_{i+1}| \leq 2e,$$

and hence that

(27.12) $$|m_n(h) - m_n(k)| \leq 2e.$$

The sum of the numerical coefficients in the series defining μ-length is 1. Relation (27.11) accordingly follows from (27.12).

The principal theorem follows:

THEOREM 27.2. The point function $q(h, \mu)$ is continuous in its arguments on the domain of admissible pairs (h, μ).

We shall prove $q(h, \mu)$ continuous at (h_0, μ_0). Let e be an arbitrary positive constant. We shall show that there exists a positive constant d such that, if (h, μ) is admissible and

(27.13) $$h \; h_0 < d \; , \quad |\mu - \mu_0| < d \; ,$$

then

(27.14) $$q(h, \mu) \; q(h_0, \mu_0) < e.$$

To that end we subject d to two conditions:

(1) Take d $<$ e/2. If h h_0 $<$ d there will exist a
homeomorphism T_d between μ-parametrizations of h and h_0
in which corresponding points have distances less than d.
If the point μ on h thereby corresponds to μ_1 on h_0,then

$$(27.15) \qquad q(h, \mu)\, q(h_0,\mu_1) < \frac{e}{2}$$

in accordance with the nature of T_d.

(2) The second condition on d is that it be so
small that when $|\mu_1 - \mu_0| < 3d$

$$(27.16) \qquad q(h_0, \mu_1)\, q(h_0, \mu_0) < \frac{e}{2}.$$

This condition can be satisfied by virtue of the continuity of $q(h_0, \mu)$ in μ.

With d so chosen, I say that (27.14) holds for admissible pairs (h, μ) satisfying (27.13). Let μ satisfy
(27.13), and let μ_1 on h_0 correspond under T_d to μ on h.
Recall that

$$(27.17) \qquad q(h, \mu)\, q(h_0, \mu_0) \leqq$$
$$q(h, \mu)\, q(h_0, \mu_1) + q(h_0, \mu_1)\, q(h_0, \mu_0).$$

The first term on the right is less than e/2 in accordance with condition (27.15) and the second likewise,pro-
vided (27.16) is applicable; that is, provided

$$(27.18) \qquad |\mu_1 - \mu_0| < 3d.$$

But under T_d a point μ on h will correspond to a point μ_1
on h_0 such that

$$| \mu - \mu_1 | \leq 2d$$

in accordance with Theorem 27.1.   Hence (27.18) holds,
(27.16) is applicable, and (27.14) holds as desired.

A particular corollary of the theorem is that when
the Fréchet distance hk = 0, then h = k.   It follows from
Theorem 27.1 that when hk = 0, $\mu_h = \mu_k$, so that when (h, μ)
is admissible (k, μ) is admissible.   It follows from
Theorem 27.2 and the condition hk = 0, that for each admissible μ

$$q(h, \mu) \, q(k, \mu) = 0 ,$$

so that

$$q(h, \mu) = q(k, \mu) \qquad (0 \leq \mu \leq \mu_h).$$

Thus h and k have a common admissible parameterization so
that h = k.   The corollary may be given the form

COROLLARY.   A necessary and sufficient condition
that hk = 0 is that h = k.

The equality h = k is, of course, the equality of
two classes of p-curves.

§28.   Admissible deformations of locally
simple curves

We must define deformations of p-curves, as distinguished from deformations of curves given as classes of
p-curves (parameterized curves).   A 1-parameter family of
sensed closed p-curves of the form

$$(28.1) \qquad u + iv = w = f(\theta, t), \qquad (t_0 \leq t \leq t_1, \ t_0 < t_1)$$

with t constant and θ variable, $0 \leqq \theta \leqq 2\pi$, on each
curve, with

$$f(\theta + 2\pi, t) \equiv f(\theta, t)$$

and $f(\theta, t)$ continuous in θ and t, and with the set of
p-curves uniformly locally simple, will be said to define
an admissible deformation of the p-curve

$$w = f(\theta, t_0) \qquad \text{into the p-curve} \qquad w = f(\theta, t_1).$$

The following lemma shows that the possibility of deforming a p-curve of a curve class g_0 into a p-curve of a
curve class g_1 is independent of the parameterizations of
g_0 and g_1 which are selected.

LEMMA 28.1. Any two representations of an admissible
curve g can be admissibly deformed into each other.

Suppose that

(28.2) $w = F(\theta)$ $(0 \leqq \theta \leqq 2\pi)$

is one representation of g, and corresponding to the
homeomorphism for which

$$\theta' = h(\theta) \qquad h(\theta + 2\pi) \equiv h(\theta) + 2\pi \qquad (0 \leqq \theta \leqq 2\pi),$$

(28.3) $w = F(h(\theta)) = H(\theta)$

is a second representation of g. For each value of t on
the interval $0 \leqq t \leqq 1$ the equation $\theta_1 = t\, h(\theta) + (1-t)\theta$
defines an admissible change of parameter from θ to θ_1,
θ_1 increasing with θ and

$$\theta_1(\theta + 2\pi) \equiv \theta_1(\theta) + 2\pi.$$

The deformation

$$w = F[t\ h(\theta) + (1 - t)\theta]$$

is admissible and deforms the p-curve (28.2) into the p-curve (28.3).

It is accordingly legitimate to refer to admissible deformations of curves as well as of p-curves.

To establish the existence of an admissible deformation of a curve g_0 into a curve g_1 one must be assured of some representation of the deformation of the form (28.1). For example, suppose one deforms a polygon by moving its vertices continuously, keeping the polygons uniformly locally simple by keeping the lengths of the edges bounded from zero and keeping any edge from intersecting an adjacent edge in more than a point. Is the resulting family of polygons representable as a deformation? It is clear that the polygons move continuously in the sense of Fréchet. Does it follow that they can be parameterized after the manner of (28.1)? Theorem 28.1 gives an answer.

We are admitting curves with representations p = p(θ) in which p(θ) is constant on no sub-interval of $0 \leqslant \theta \leqslant 2\pi$. A 1-parameter family g_t ($0 \leqslant t \leqslant 1$) of such curves will not in general be given with a representation of the form (28.1). We admit the possibility that g_t depend upon t in no continuous manner. With this understood the following theorem is relevant.

THEOREM 28.1. Let g_t ($0 \leqslant t \leqslant 1$) be a 1-parameter family of admissible, sensed closed curves, with t constant on each curve. A necessary and sufficient condition that the family g_t admit a representation

$$w = f(\theta, t) \qquad\qquad (0 \leqq \theta \leqq 2\pi;\ 0 \leqq t \leqq 1)$$

in which f(θ, t) is continuous in θ and t and has the period 2π in θ, <u>is that there exist a point</u> Q_t <u>on</u> g_t <u>such</u> <u>that the arc</u> g_t^* <u>obtained by cutting</u> g_t <u>at</u> Q_t <u>vary contin-</u> <u>uously with</u> t <u>in the sense of Fréchet</u>.

That the condition is necessary is immediately obvious on setting $Q_t = f(0, t)$.

To prove the condition sufficient we refer each arc
g_t^* to its μ-length, measuring the μ-length from Q_t and
setting

$$\mu = \frac{\theta \mu_t}{2\pi},$$

where μ_t is the total μ-length of the curve g_t^*. The resulting representation $f(\theta, t)$ of the curves g_t will have
the required form in accordance with Theorem 27.2, at
least for $0 \leq \theta \leq 2\pi$. But $f(0, t) \equiv f(2\pi, t)$, and $f(\theta, t)$
can accordingly be extended in definition so as to have
the period 2π in θ. The theorem follows.

By a <u>broken analytic curve</u> will be meant a finite
sequence of regular analytic arcs. An analytic arc is
representable in the form

$$x = x(\theta) \qquad y = y(\theta),$$

where $x(\theta)$ and $y(\theta)$ are real analytic functions. Neighboring a corner P of a broken analytic curve we shall
suppose that the two arcs intersecting at P, intersect only
at P. Two closed analytic arcs intersect either in a
finite number of points, not at all, or in a sub-arc of
the arcs. Any locally simple closed curve can be admissibly deformed into a broken analytic curve or a regular
curve, as we shall see. We begin with a lemma.

LEMMA 28.2. <u>Let</u> a <u>be a simple arc and</u> b <u>a sub-arc</u>
<u>of</u> a <u>whose end points</u> B_1 <u>and</u> B_2 <u>are inner points of</u> a.
<u>There exists an open, simple, regular, analytic arc</u> c

whose end points are B_1B_2 and which intersects a in B_1 and B_2 only. If a is regular and analytic at B_1 (or B_2), the arc c may be chosen so as to be regular and analytic at B_1 (or B_2). There exists an admissible deformation of b into c in which B_1 and B_2 are fixed and b is deformed through simple arcs which intersent a only in B_1 and B_2. The arc c and the deformation of c into b can be taken on an arbitrarily small neighborhood of b.

     To prove the lemma let g be a simple closed curve of which a is a sub-arc. Let the finite region bounded by g be mapped in a 1 - 1 and directly conformal manner onto the interior R of a unit circle. In this mapping let b' be the image of b on the unit circle. Let k' be a circular arc joining the end points of b' within R, and let k be the antecedent of k' under the conformal transformation T. It is clear that k can serve as the arc c of the lemma. If k' is deformed into b' through a family of circular arcs joining the end points of b', the antecedents of these circular arcs under T will define a deformation of the kind required in the lemma. If the arc a is regular and analytic at B_1 (or B_2) the transformation T is directly conformal with a non-vanishing Jacobian at B_1 (or B_2). See Osgood (p. 719). The arc k is accordingly analytic in some neighborhood of B_1 (or B_2).

     The lemma follows.

     In proving that any locally simple closed curve g can be admissibly deformed into a broken analytic curve arbitrarily near g in the sense of Frechet, use will be made of a representation of g as a circular sequence

(28.4) $a_1, a_2, \ldots, a_n$ $(n > 2)$

of arcs any successive three of which form a simple sub-arc of g. Such a sequence will be called an n-sequence. The minimum distance d_n between any two successive ver-

tices (end points of arcs of the sequence) of (28.4) will
be called the vertex norm of the sequence. A vertex norm
d_n is a norm of local simplicity of g; for any sub-arc of
g whose diameter is less than d_n is a sub-arc of some sequence of two successive arcs of (28.4) and hence simple.

The curve g will be continuously deformed through a
family g^t, $0 \leq t \leq 1$, of n-sequences. It is understood
that the vertices of the n-sequence g are deformed through
the vertices of g^t. The minimum of the vertex norms of
the n-sequence g^t, for $0 \leq t \leq 1$, will be a positive number which is a norm of local simplicity of the curves of
the deformation g^t.

The concept of an n-sequence and its vertex norm per-
mits the enunciation of the following useful lemma.

LEMMA 28.3. Let g be a locally simple closed curve
representable as an n-sequence with a vertex norm 4e.
For any set Ω of locally simple curves which are representable as n-sequences with successive vertices within a
distance e of the respective vertices of g, 2e is a norm
of local simplicity.

This follows from the fact that the minimum distance
between successive vertices of the given n-sequences exceeds 2e.

We come to a major theorem.

THEOREM 28.2. Any locally simple, closed curve g can
be admissibly deformed on an arbitrarily small Fréchet
neighborhood of g into a broken, analytic curve g',
through a family of curves whose norm of local simplicity
is independent of the nearness of the family to g.

Let g be represented by the n-sequence (28.4) with a
vertex norm of 4e. The required deformation will be
through a family of n-sequences, of which the initial sequence will be the sequence (28.4). In accordance with

the preceding lemma the family will have a norm of local
simplicity 2e, provided the deformation of the vertices
of (28.4) displaces each vertex by a distance less than e.

By virtue of Lemma 28.2, g can be admissibly deformed
into a closed curve g_1 in such a manner that a_1 is deformed into a simple arc a_1' joining the end points of a_1,
analytic and regular at its interior points, while the
remainder of g remains fixed. We can suppose that the
deformation of a_1 has been through arcs so near a_1 that

(28.5) $\qquad\qquad a_1', a_2, \ldots, a_n$

is an n-sequence, as well as its antecedents in the deformation of g into g_1.

Without changing g_1 as a whole, or altering $a_3, \ldots,$
a_n, the arcs a_1' and a_2 of (28.4) will be slightly shorten-
ed and extended respectively on g_1 near their common end
point. Let a_1'' and a_2' be the arcs thereby replacing a_1'
and a_2. By virtue of Lemma 28.2, the curve g_1 can be admissibly deformed into a closed curve g_2 in such a manner
that a_2' is replaced by a simple arc a_2'' which is regular
and analytic in its interior and at its initial end point,
that the remaining arcs of g_1 remain fixed, and that,

(28.6) $\qquad\qquad a_1'', a_2'', a_3, \ldots, a_n$

is an n-sequence, as well as its antecedents in the deformation of g_1 into g_2.

One then operates on a_2'' and a_3 in (28.6) as upon a_1'
and a_2 in (28.5), obtaining a new n-sequence

(28.7) $\qquad\qquad a_1'', a_2''', a_3'', a_4, \ldots, a_n,$

and then upon a_3'', a_4 in (28.7), and so on until an n-sequence

$$a_1'', \ a_2''', \ a_3''', \ \ldots, \ a_{n-2}''', \ a_{n-1}'', \ a_n$$

is reached. The arc a_n is then slightly enlarged at both ends at the expense of a_1'' and a_{n-1}'', and deformed as above into a simple regular analytic arc a_n''. The resulting broken analytic curve g' will satisfy the theorem, provided merely that the above deformations have been on a sufficiently small Fréchet neighborhood of g and the vertices of g in (28.4) have been displaced by distances less than e. That this is possible is seen from Lemma 28.2. In accordance with Lemma 28.3 a norm of local simplicity of the family of n-sequences through which g is deformed is then 2e.

The proof of the theorem is complete.

THEOREM 28.3. Any admissible broken analytic curve g can be admissibly deformed into a regular curve.

The theorem will be proved by removing the corners of g by local deformations.

To that end let A be a corner of g, and let h be a simple arc of g which contains A as an inner point but which is otherwise without singularity. With A as a center let C be a circle of so small a radius that C intersects h in just two inner points M and N. The arcs AM and NA of h together with a circular arc of C will form a curvilinear triangle D which together with its interior can be mapped homeomorphically onto the interior and boundary of an isosceles triangle D', conformally at all interior points and with vertices corresponding to vertices.

Let A' be the image of A on D'. Let K denote the circle inscribed in D', and let k be that sub-arc of K

whose end points are on the sides of D' incident with A'
and which is nearest A'. Let k_1 be the sub-arc of D'
which contains A' and has the end points of k. The arc k
can be admissibly deformed into the arc k_1 on the domain
bounded by k and k_1. The inverse of the above conformal
map transforms this deformation into an admissible deformation of h into a regular arc.

It is clear that the above deformation of g can be
made on an arbitrarily small Fréchet neighborhood of g.
The following theorem is needed.

THEOREM 28.4.   Any closed curve g with a regular representation

$$w = u(\theta) + i\, v(\theta) \qquad (0 \leqq \theta \leqq 2\pi)$$

can be admissibly deformed on an arbitrarily small Fréchet
neighborhood of g into a regular, analytic, closed curve.

Let (r, θ) represent polar coordinates in the plane
of a complex parameter $z = x + i\, y$. Let U(r, θ) and
V(r, θ) be functions which are harmonic in the coordinates
(x, y) for $r < 1$, continuous for $r \leqq 1$, and satisfy the
conditions

$$U(1, \theta) \equiv u(\theta) \qquad\qquad V(1, \theta) \equiv v(\theta).$$

Such functions are given by the Poisson integral with
boundary values u(θ), and v(θ) respectively. Since u'(θ)
and v'(θ) exist and are continuous, $U_\theta(r, \theta)$ and $V_\theta(r, \theta)$
exist and are continuous for $r \leqq 1$, as is well known. If
$r_0 < 1$ is any constant sufficiently near 1, the family of
curves (r constant) for which

$$w = U(r, \theta) + i\, V(r, \theta) \qquad (r_0 \leqq r \leqq 1)$$

are regular for $r_0 \leq r \leq 1$ and analytic for $r_0 \leq r < 1$.
This family will then admissibly deform g into the curve
of the family for which $r = r_0$ in accordance with the re-
quirements of the theorem.

§29. Deformation classes of locally simple curves.

We have seen that any locally simple, closed curve
can be admissibly deformed into a regular curve. It has
also been seen that two locally simple curves which can
be admissibly deformed into each other must have the same
angular order p. We shall establish the converse of this
statement.

For the special case of regular curves Theorem 29.1
has been proved by Graustein and Whitney. See Whitney
(2). The interpolation procedure of Whitney is used here
with simplifications. In particular, our proof does not
distinguish between the cases p = 0 and p ≠ 0. The proof
by Morse and Heins (1), I, is combinatorial in character
and suggests the possibility of generalization to the
case n > 2. The following lemma is a necessary preliminary to the theorem.

LEMMA 29.1. Let m(s) be a continuous, non-constant,
complex function of the real variable s on the interval,
$0 \leq s \leq 1$ with $|m(s)| = 1$. Then the complex number

$$(29.1) \qquad\qquad K = \int_0^1 m(s)\, d\, s$$

has an absolute value $|K| < 1$.

If K = 0 the lemma is true. If K ≠ 0, let $\alpha = $ arc K.
Then $K\, e^{-i\alpha}$ is the real number

$$K^* = \int_0^1 R[e^{-i\alpha}m(s)]\ ds.$$

Thus K* is the real average of a real continuous function whose absolute value is not constant and is at most 1. Hence $|K^*| < 1$ and the lemma follows.

THEOREM 29.1. Any two locally simple, sensed closed curves with the same angular order can be admissibly deformed into each other.

By virtue of the results of the preceding section it will be sufficient to prove the theorem for two regular curves g_1 and g_2 with the same angular order p. A continuous movement of g_1 or g_2 as rigid bodies is clearly an admissible deformation, as is a continuous family of similarity transformations of g_1 or g_2. We can accordingly suppose that g_1 and g_2 have the same total length 1, and that the point s = 0 on both curves is the point z = 0 in the complex z-plane, while the positive tangent to both curves at s = 0 is the positive x-axis.

With this understood suppose that g_1 and g_2 have the respective representations

$$z = h(s) \qquad z = k(s) \qquad (0 \leq s \leq 1)$$

in terms of their arc length s, and that

(29.2) $h(0) = k(0) = h(1) = k(1) = 0$

(29.3) $h'(0) = k'(0) = 1.$

The complex numbers h'(s), k'(s) have unit absolute values and so can be represented as points on a circle C of unit radius with center at the origin of a complex plane. Let

C' be the unending simply-connected 1-manifold which
covers C. Let H_s be a sub-arc of C' which leads from a
point covering h'(s) to a point covering k'(s), which varies continuously on C' with s, and whose length reduces to
zero when s = 0. Such a sub-arc H_s is uniquely determined
for each value of s. For each value of the time t on the
interval (0, 1), let m(t, s) be a complex point on H_s
which divides H_s, with respect to length, in the ratio of
t to 1 - t. For each s

(29.4) $m(0, s) \equiv h'(s)$ $m(1, s) = k'(s).$

Without loss of generality we can suppose that for
some small positive constant e and for $0 < s < e$

(29.5) $0 < \text{arc}^* h'(s) < 2\pi,$ $0 < \text{arc}^* k'(s) < 2\pi.$

Since h'(0) = k'(0) = 1, (29.5) can be made to hold by
subjecting g_1 and g_2 to a suitable small admissible deformation affecting points near s = 0 only. Relations
(29.2) and (29.3) are to remain valid.

The proof proper begins at this point. The deformation of g_1 into g_2 will be defined by the family of curves

(29.6) $z = f(t, s) = \int_0^s [m(t, s) - K(t)]\, ds$

where K(t) will be determined as the average

(29.7) $K(t) = \int_0^1 m(t, s)\, ds$

of m(t, s) with respect to s. In accordance with (29.5)

$0 < \text{arc}^* m(t, s) < 2\pi$ $(0 < s < e),$

----
*For a suitable choice of the arc.

and since $m(t, 0) \equiv 1$, $m(t, s)$ is identically constant as a function of s for no constant t on $(0, 1)$. It follows from the preceding lemma that $|K(t)| < 1$.

The curves g_t of the family are closed. This is a consequence of the identities

$$f(t, 0) \equiv 0 \qquad\qquad f(t, 1) \equiv 0.$$

The second identity holds by virtue of the choice of $K(t)$.
• The curves of the family are regular. To see this note that

$$(29.8) \qquad\qquad z_s = m(t, s) - K(t) \neq 0$$

since $|K| < 1$. The derivative z_s is a vector tangent to g_t at each point of g_t, with the possible exception of the junction point at which $s = 0$ and 1. At this junction point (29.8) gives a value of 1 for z_s, both when $s = 0$ and $s = 1$, since

$$(29.9) \quad m(t, 0) \equiv m(t, 1) \equiv 1 \qquad K(0) = K(1) = 0.$$

We shall verify the relations (29.9).

First $m(t, 0) \equiv 1$, since the arc H_s reduces to the point 1 when $s = 0$. When $s = 1$, H_s similarly reduces to the point 1 as a consequence of the hypothesis that g_1 and g_2 have the same angular order. Hence $m(t, 1) \equiv 1$. Finally

$$K(0) = \int_0^1 m(0, s)\, ds = \int_0^1 h'(s)\, ds = 0,$$

since $h(s)$ represents a closed curve. Similarly $K(1) = 0$, since $k(s)$ represents a closed curve. Thus (29.9) holds and g_t has no corner at $s = 0$ and 1. Thus g_t is regular without exception.

It follows from (29.6) that

$$f(0, s) \equiv h(s) \qquad\qquad f(1, s) \equiv k(s)$$

so that $f(t, s)$ admissibly deforms g_1 into g_2. The proof of the theorem is accordingly complete.

A curve C^n of the form

$$z = e^{in\theta} \qquad\qquad (n = \pm 1, \pm 2, \ldots, \quad 0 \leqq \theta \leqq 2\pi)$$

has the angular order n. The curve C^o given by

$$z = \sin 2\theta + i \sin \theta \qquad (0 \leqq \theta \leqq 2\pi)$$

is a figure eight of angular order 0. One has the following theorem.

THEOREM 29.2. A locally simple, sensed, closed curve with the angular order p is admissibly deformable into the canonical curve C^m if and only if $p = m$.

§30. The product of locally simple curves.

The notion of the product of two locally simple curves g_1 and g_2 will be defined under certain conditions (A) which we shall specify.

Let g_1 and g_2 be represented in terms of a parameter u in a locally 1 - 1 manner with a period ω in u. Suppose that g_1 and g_2 intersect in a point Q represented by $u = 0$ on both curves. By the cross join g of g_1 and g_2 at $u = 0$ will be meant the curve obtained by tracing g_1 from $u = 0$ to $u = \omega$, then g_2 from $u = 0$ to $u = \omega$ identifying

the point $u = \omega$ on g_2 with the point $u = 0$ on g_1.
In order that the join g be locally simple it is
necessary and sufficient that there exist a simple arc h_1
of g_1 containing $u = 0$ in its interior and a simple arc
h_2 of g_2 containing $u = 0$ in its interior, such that the
two composite arcs of g obtained from h_1 and h_2 are simple.
Suppose that $u = 0$ divides h_1 into arcs a' and b' with a'
preceding $u = 0$. Similarly suppose $u = 0$ divides h_2 into
arcs a" and b", with a" preceding $u = 0$. We term a' and
a" incoming arcs and b' and·b" outgoing arcs. See Fig. 5.
If then the arcs h_1 and h_2 are sufficiently limited an incoming arc intersects neither outgoing arc except at Q,
and an outgoing arc intersects neither incoming arc except at Q, provided g is locally simple.

(A) We shall assume that g_1, g_2 and the cross join g
of g_1 and g_2 are locally simple, and that there exists a
simple arc k through Q which,in a sufficiently small
neighborhood of Q intersects sufficiently limited incoming
and outgoing arcs of g_1 and g_2 only in Q,and separates the
incoming arcs from the outgoing arcs.

We shall denote the angular order of a locally simple
curve r by p(r) and prove the following theorem.

THEOREM 30.1. When (A) holds, then

(30.1) $p(g) = p(g_1) + p(g_2)$.

We state without proof the fact that when g_1 and g_2
and the cross join g are locally simple and (A) does not

hold, then the incoming and outgoing arcs, if sufficiently
limited, intersect only at Q with the incoming arcs sep-
arating the outgoing arcs. This case can be described as
one in which g_1 and g_2 contact with contrary senses, and
it can be shown that

$$(30.2) \qquad p(g) = p(g_1) + p(g_2) \pm 1.$$

We make no use of this relation.

    Proof of 30.1. Relation (30.1) is obvious in the
special case in which the two incoming arcs are identical,
and the two outgoing arcs are identical. The proof of
(30.1) consists in showing that this special case arises
after a preliminary admissible deformation of g_1, g_2, and
g neighboring the point Q with parameter u = 0. The
point Q is held fast during this deformation.

    Let C be a circle with center at Q so small that
both of the incoming arcs and outgoing arcs in (A) intersect C, while the arc k in (A) intersects C both before
and after its intersection with Q. Without loss of generality we can suppose that the arcs a', a", b', b", each
intersect C in one end point, and that k is a simple arc
with end points on C but otherwise within C. See Fig. 5.

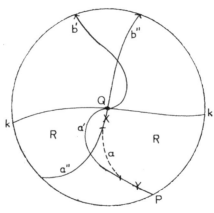

Figure 5.

Let D be the disc bounded by C and let R be the
(open) subregion of D bounded by k and containing the inner points of a' and a". Let a be a simple open arc
which lies on R and which is identical with a" on some
terminal sub-arc X of a", and with a' on some initial sub-
arc Y of a'. It is clear that a can be deformed on R
into the open arc of a' through a family of simple arcs
on R each of which has Y as an initial sub-arc and Q as a
terminal point. This deformation can be regarded as an
admissible deformation of g_1 and of g. The fact that Y
is unmoved insures that the moving arcs do not intersect
their complements on g_1 and g so as to violate the condition of uniform local simplicity. We can accordingly
suppose that the incoming arcs on g_1 and g_2 are identical
on some sub-arc X terminating at Q. Similarly, we can
suppose that the outgoing arcs are identical on some sub-
arc beginning at Q. For a situation of this sort, relation (30.1) is obvious, and the proof of the theorem is
complete.

§31. The product of deformation classes.

The product $g_1 g_2$ of two locally simple, sensed,
closed curves has been defined under the following circum-
stances.

(1) Suitably parameterized the curves g_1 and g_2 intersect in a point Q represented on both curves by a parameter value u = 0.

(2) The "cross join" of g_1 and g_2 at u = 0 is locally simple.

(3) There exists a simple arc k with Q as an inner
point such that sufficiently limited simple arcs of g_1 and
g_2, preceding u = 0 on g_1 and g_2 respectively, are separ-
ated by k on some neighborhoods of Q from sufficiently
limited arcs of g_1 and g_2 following u = 0, and intersect k
only in Q.

When (1), (2) and (3) are satisfied the "cross join"
of g_1 and g_2 is termed the product g_1g_2 of g_1 and g_2. We
have seen that

(31.1) $p(g_1g_2) = p(g_1) + p(g_2)$.

Let the class of all locally simple closed curves
which are admissibly deformable into a locally simple
closed curve h be denoted by [h]. We term [h] a deformation class. All curves of [h] have the same angular
order. Given two deformation classes

$$c_1 = [h_1] \qquad c_2 = [h_2]$$

the product c_1c_2 will be defined: Let g_1 and g_2 be
curves of $[h_1]$ and $[h_2]$ respectively for which the product is defined by a "cross join" of g_1 and g_2 at some
point Q. Curves g_1 and g_2 for which g_1g_2 is defined for
some point of intersection Q always exist. One can in
particular take g_1 and g_2 as regular representations of
$[h_1]$ and $[h_2]$ tangent in the positive sense to each other
at a point of intersection Q. The conditions for the existence of g_1g_2 are then satisfied.

We define the product c_1c_2 as the deformation class
$[g_1g_2]$,and show that this deformation class is independent of the choice of g_1 and g_2 among curves of $[h_1]$ and
$[h_2]$ for which g_1g_2 is defined.

Let g_1' and g_2' be two other curves of $[h_1]$ and $[h_2]$
respectively for which $g_1'g_2'$ is defined. Then

$$p(g_1) = p(g_1'), \qquad p(g_2) = p(g_2').$$

It follows from (31.1) that

$$p(g_1 g_2) = p(g_1' g_2').$$

We make use of the theorem that any two locally simple curves with the same angular order can be admissibly deformed into each other, and infer that

$$[g_1 g_2] = [g_1' g_2'].$$

The product $c_1 c_2$ is thus uniquely defined as a deformation class.

Let the angular order of the curves of a deformation class c be denoted by $P(c)$. Previous results can be summarized in the following lemma.

LEMMA 31.1. There is a 1 - I correspondence between the multiplicative domain of deformation classes c and the integers p, in which c corresponds to $P(c)$ and

(31.2) $P(c_1 c_2) = P(c_1) + P(c_2).$

As a consequence of this lemma one can prove the following theorem.

THEOREM 31.1.* The deformation classes c form an abelian group G isomorphic with the additive group J of integers in which c corresponds to $P(c)$.

Commutativity in G. That $c_1 c_2 = c_2 c_1$ follows from the fact that

---

*This theorem is an abstract consequence of Lemma 31.1 independent of the interpretation of c as a deformation class. The proof which we give is in reality a proof of this abstract principle.

$$P(c_1 c_2) = P(c_2 c_1)$$

in accordance with (31.1),and the fact that there is but one class c with a given P.

Associativity. That

(31.3.) $$c_1(c_2 c_3) = (c_1 c_2)c_3$$

follows from the identity of the angular order P of the two sides of (31.1).

The unit element e.   We define e in G as the class e for which P(e) = 0. That

$$ce = ec = c$$

follows from the equality of the angular order of ce, ec, and e.

The inverse c^{-1}.   We define c^{-1} as the class whose angular order is -P(c). That

(31.4) $$cc^{-1} = c^{-1}c = \epsilon$$

then follows from the equality of the angular order of the terms in (31.4).

The proof of the theorem is complete.

We note that e is the deformation class of the figure eight. When [g] is a given deformation class the inverse class is $[g^{-1}]$ where g^{-1} is g reversed in sense. This does not follow from the nature of gg^{-1}, for in fact this product is not defined, but rather from the fact that the sum of the angular orders of g and g^{-1} is zero.

Certain special relations are of interest. If g is a

sensed regular curve g^n will be defined for $n > 0$ as g
traced n times in the positive sense, and for $n < 0$, as
g traced $-n$ times in the negative sense. Let g' be the
reflection of g in the tangent to g at a point $u = 0$ on
g. The product gg' will be denoted by g^0. We have the
relation

$$(31.5) \qquad [g^n] \, [g^m] = [g^{n+m}] \; ,$$

where n and m are arbitrary integers. This follows from
the equality of the angular order of the classes on the
two sides of (31.5).

§32. 0-Deformations. Curves of order zero.

An admissible deformation none of whose curves intersect the origin 0 will be called an 0-deformation.

The previous admissible deformations in the w-plane
have kept the angular order p invariant, but have permit-
ted deformations through the point $w = 0$. The order q of
the given curve with respect to $w = 0$ could accordingly
change. In the case in which one is deforming an interior transformation $w = f(z)$, one does not wish to admit
new zeros during the deformation. Suppose, for example,
that $f(z)$ is defined on a region G bounded by a single
Jordan curve B with a locally simple image g under $w =
f(z)$. If one wishes to avoid the introduction of new
zeros of $f(z)$ one should deform g so that the moving
curve g_t never intersects $w = 0$. If one uses 0-deforma-
tions this will be accomplished, both p and q remaining
invariant.

We accordingly seek canonical curves under 0-deforma-
tions operating on a locally simple curve g which does
not intersect the point $w = 0$. In the important problem
of finding canonical forms for interior transformations

with a prescribed number of fixed zeros, poles, and
branch point antecedents a first step is to get canonical
curves for the boundary images. The deformations to
which the interior transformatiuns will be subjected will,
in their effect on the boundary images, be 0-deformations.
The classification of boundary images in deformation types
under 0-deformations will not be fine enough for the ulti-
mate homotopic classification of interior transformations.
It is a study of first necessary conditions, a division
into classes which must be still further subdivided in
later developments. However the canonical curves under
0-deformations correspond in a one-to-one way to the pairs
of integers (p, q) as angular order and order of g. These
canonical curves reveal the meaning of the pairs (p, q).

   We start with the case in which the order q = 0, and
prove the following lemma.

   LEMMA 32.1. Let g be a locally simple, regular curve
which does not intersect w = 0 and whose order q = 0. Let
R be a point of g and E a line element tangent to g at R.
There exists a regular 0-deformation of g onto an arbitrary
neighborhood of R leaving R and E unchanged during the
deformation.

   Let (r, θ) be polar coordinates in the w-plane. Let
g be referred to arc length s as a parameter, with b the
total length of g. Without loss of generality we can suppose that R is the point w = 1. The curve g admits a representation of the form

$$r = r(s) \qquad\qquad \theta = \theta(s) \qquad (0 \leq s \leq b)$$

in which r(s) and θ(s) are continuous in s. Moreover,
θ(0) = θ(b) by virtue of the fact that q = 0 on g. We
can suppose that s = 0 at R and that θ(0) = 0.

   In the deformation D which we shall define, the time

t shall run from 0 to 1 - e, where e is an arbitrarily
small positive constant. Under D the point w on g shall
be replaced by the point w^{1-t} where the particular branch
of w^{1-t} to be used is defined as follows. The polar coordinates of w^{1-t} shall be

$$[r(s)]^{1-t}, \quad (1-t) \, \theta(s) \qquad (0 \leqq t \leqq 1-e)$$

thus giving the image at the time t of the point at
$[r(s), \theta(s)]$ when t = 0. The point R is represented by
s = 0, with $(r, \theta) = (1, 0)$, and is fixed under D. The
transformation from w to w^{1-t} for a fixed t is conformal
at w = 1, that is, at R. The branch which we have used
near w = 1 carries the direction $\theta = 0$ into itself. It
follows from the conformality that the direction E at R
is unchanged during D.

The final image of g under D when t = 1 - e will consist of the points (r_1, θ_1) for which

$$r_1 = r^e(s) \qquad \theta_1 = e\theta(s) \, ,$$

and will be on an arbitrary neighborhood of the point r =
1, $\theta = 0$, if e is sufficiently small.

The proof of the lemma is complete.

We can now prove the following theorem.

THEOREM 32.1. Let g be a locally simple sensed curve
which does not intersect the origin and whose order q = 0.
If p is the angular order of g there exists an 0-deforma-
tion into a canonical curve C^p where C^p is a positively
sensed circle C, traced p times when $p \neq 0$ without encircling the origin, and C^o is a figure eight of which neither loop encircles the origin.

Without loss of generality we can suppose that g is

regular, since there exists an admissible deformation of
g into a regular curve, and since this deformation can be
chosen so as to displace the points g by so small distances that the point w = 0 is not reached. With g regular, one can apply the preceding lemma and so 0-deform g
into a regular curve g_1 on an arbitrarily small neighborhood of a point R on g. Let T_1 be an admissible deformation of g_1 (Cf. Theorem 29.2) into a curve of the deformation type specified in the lemma. T_1 may not be an 0-def-
ormation. But if one first sufficiently contracts g_1 and
its images under T_1, towards R as a center of similarity,
the curve g_2 into which g_1 is contracted will be 0-def-
ormed, under the deformation T_2 into which T_1 is contracted, into the required canonical form. A preliminary
deformation contracting g_1 into g_2 is of course necessary,
and we can suppose that g_1 is on so small a neighborhood
of R that this contraction is an 0-deformation.

    The Theorem follows.

    Note. It is of importance for a proof of Theorem
33.1 that the final deformation T_2 can be taken as one
which holds fast the point R of g and the element E tangent to g at R.

    §33. 0-Deformations. Curves of order q $\neq$ 0.

    Let g be a locally simple curve which does not inter-
sect w = 0. We shall be concerned with canonical forms
for curves such as g under 0-deformations. No generality
will accordingly be lost if g is replaced by a regular
curve into which g may be 0-deformed. We accordingly suppose that g is regular.

    Let (r, θ) be polar coordinates in the w-plane. The
curve g lies on an annulus

$$r_1 < r < r_2 \qquad\qquad (r_1 > 0)$$

in the w-plane. It will be convenient to represent this

annulus by a strip M in a plane of rectangular coordinates (r, θ). This strip is bounded by the straight lines $r = r_1$ and $r = r_2$. If $w = u + iv$ we have

$$u = r \cos \theta \qquad v = r \sin \theta.$$

Let g be referred to its arc length s, with b the total length of g. Let $R(s)$ be an image point on M of the point s on g, with $R(s)$ defined and continuous for $0 \leq s \leq b$. Let $\theta(s)$ be the value of θ at $R(s)$. We have

$$\theta(b) - \theta(0) = 2q\pi$$

where q is the order of g.
We shall prove the following lemma.

LEMMA 33.1. If $q \neq 0$ the curve g can be 0-deformed into a product

$$(33.1) \qquad\qquad C^q X \,,$$

where C is a circle with center at $w = 0$, and X is a sensed regular curve positively tangent to C^q at a point Q at which C^q is cross-joined to X to define the product $C^q X$. The order of X is 0, and the angular order is p - q.

As previously we suppose that g is regular and represented on the strip M by an arc Z defined by $R(s)$ on M, with $0 \leq s \leq b$. See Fig. 6. There is at least one point $s = a$ on Z at which the value of $r(s)$ on Z is an absolute minimum r_0. With proper choice of $s = 0$ on g, a differs from 0 and b. We can suppose that $r(s) = r_0$ at no point of Z other than $s = a$. A suitable 0-deformation of g which moves points of g near the point $s = a$ will bring this about. More precisely we can suppose that for values of s for which

$$(33.2) \qquad\qquad a - 2e \leqq s \leqq a + 2e \qquad\qquad (e > 0)$$

(where e is an arbitrarily small constant) Z is a para-
bolic arc Z' on M with vertex at R(a) and with vertical
axis. Let Z" be the complement of Z' on Z. We can suppose Z' such that on Z', r(s) is less than its value at
any point on Z". The unbroken curve in Fig. 6 represents
Z.

To continue we shall suppose that $q > 0$. If $q < 0$
one could reverse the sense of g and obtain a new curve g'
for which $q > 0$. The truth of the Lemma for g' implies
its truth for g.

We shall deform the sub-arc

$$(33.3) \qquad\qquad a \leqq s \leqq a + 2e$$

of Z, keeping the remainder of Z fixed. The time t shall
vary on the interval $0 \leqq t \leqq 1$. We cut Z at $s = a$ and
move the sub-arc

$$a \leqq s \leqq a + e$$

of Z to the right with a θ-speed of 2qπ ($0 \leqq t \leqq 1$). We
fill in the gap between the point R(a) and the moving arc
by a straight line segment on $r = r_0$. The moving sub-arc
of Z finally reaches a position 2qπ units to the right of
its initial position. The points on the sub-arc

$$(33.4) \qquad\qquad a + e \leqq s \leqq a + 2e$$

will be moved to the right through variable distances with
each point at a constant r level, holding the point

$R(a + 2e)$ fast and moving $R(a + e)$ as stated above; the movement of the sub-arc defined by (33.4) can be made so that the deformation of g is regular. The terminal image of the arc (33.3) is shown by a dotted line in Fig. 6.

This deformation, made in the (r, θ) plane, is carried back to the w-plane. The horizontal line on $r = r_0$, leading to the right $2q\pi$ units from $s = a$ (Fig. 6), yields the curve C^q of the lemma. The residual curve X has the order zero, since the total order of g is invariant under an 0-deformation, and the order of a product

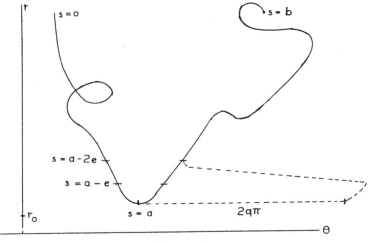

Figure 6.

of two curves is clearly the sum of the orders of the factors. The curve X is also regular and tangent to C^q at the point of g originally represented by $s = a$. The angular order of X must be $p - q$ since the angular orders of the factors must sum to p.

The proof of the lemma is complete.

The principal theorem is as follows.

THEOREM 33.1. _If_ g _is locally simple and has an order_ $q \neq 0$, _then_ g _can be_ 0-_deformed when_ $p - q \neq 0$ _into a product_

$$(33.5) \qquad\qquad C^q \, C_1^{p-q} \qquad\qquad (p - q \neq 0)$$

in which C is a positively sensed circle with center at the origin and C_1 is a circle tangent* to C but not encircling the origin. When p - q = 0, the canonical curve can be taken as C^q. When p - q ≠ 0, C_1 is internally or externally tangent to C according as q and p - q do or do not have the same sign.

To prove this theorem we start with the product C^q X of Lemma 33.1. Let E be the line element tangent to X at the point R at which X is cross joined to C^q. The curve X can be regularly 0-deformed (Cf. Theorem 32.1 and Note) through a family X_t of regular curves with fixed R and E into a curve

$$C_1^{p-q} \qquad\qquad (p - q \neq 0)$$

(of the required type) when p - q ≠ 0, and into a figure eight when p - q = 0. In the case p - q = 0 neither loop of the figure eight can encircle the origin, since the order of X_t is constantly 0.

In the case in which q = 1 it is easy to see that the product of C and the figure eight is 0-deformable into C. In this deformation one moves only the figure eight and a small arc of C neighboring the point of cross junction. This same 0-deformation can be applied when C is replaced by C^q. The canonical curve thus reduces to C^q when p - q = 0.

The proof of the theorem is complete.

COROLLARY 33.1.   If g_1 and g_2 are two locally simple curves with the same order q and angular order p, then g_1

---

*Both C and C_1 are taken counter clockwise. The sense of Cq depends on the sign of q. Similarly with a power of C_1.

and g_2 are mutually 0-deformable into each other.

This follows from the fact that g_1 and g_2 are 0-deformable into the same canonical curves. Locally simple curves g in the w-plane which do not intersect w = 0 and which are mutually 0-deformable into each other form a set termed an 0-deformation class c, or simply an 0-class. The order q and angular order p of curves of c are the same and may be denoted by Q(c) and P(c) respectively. In order that two 0-classes c_1 and c_2 be identical it is necessary and sufficient (Corollary 33.1) that

$$P(c_1) = P(c_2) \qquad Q(c_1) = Q(c_2).$$

The canonical forms show that there is an 0-class with prescribed orders (P, Q).

There is accordingly a 1 - 1 correspondence between 0-classes c and pairs of integers (P, Q) in which c corresponds to [P(c) Q(c)].

Let c_1 and c_2 be two 0-classes. The product $c_1 c_2$ can be defined as the 0-class of the product $g_1 g_2$ of any two curves g_1 of c_1 and g_2 of c_2 for which $g_1 g_2$ is defined. Given c_1 and c_2 the 0-class $c_1 c_2$ so defined is unique, since the pair

$$P(c) = P(c_1) + P(c_2)$$
$$Q(c) = Q(c_1) + Q(c_2)$$

is unique. The orders (P, Q) of c_1 and c_2 are thus added like vectors to obtain the orders of $c_1 c_2$. As in §31 we can then establish the following theorem:

THEOREM 33.2. The 0-deformation classes c with the above definition of product, form an abelian group G iso-

morphic to the additive group of pairs of integers (P, Q),
with c corresponding to [P(c), Q(c)]. The unit element
in G is the 0-deformation class of a figure eight neither
loop of which encircles the origin.

§34. Deformation classes of meromorphic
functions and of interior transformations

     In this last section a brief introduction will be
given to a deformation theory under which the theory of
functions of a complex variable presents a new aspect.
No proofs will be given.

     We shall consider interior transformations·f from
the open disc S = [|z| < 1] to the complex w-sphere, in
which f has a finite set of zeros,

$$a_0, a_1, \ldots, a_r \qquad (r > 0),$$

a finite set of poles,

$$a_{r+1}, \ldots, a_n \qquad (n > 1),$$

and branch point antecedents ,

$$b_1, \ldots, b_\mu \qquad (\mu \geq 0).$$

We shall assume that the zeros, poles and branch points
have the multiplicities 1. The case in which these multiplicities exceed 1 can be similarly treated provided that
the orders remain constant during the proposed deformations of f. The case in which n = 1 or 0 is exceptional
and will be omitted in this introduction. These exceptional cases offer no difficulties. We are also supposing
that there is at least one zero. In the case in which
there are no zeros, but poles, one can replace f by its

reciprocal. The zeros, poles, and branch point antecedents will form an ordered set of points

(34.1) $(\alpha) = (a_0, a_1, \ldots, a_n, b_1, \ldots, b_\mu)$

which will be termed a characteristic set.
     Admissible f-deformations. We shall admit deforma-
tions D of f of the form

(34.2) $w = F(z, t)$ $(|z| < 1,\ 0 \leqq t \leqq 1)$

with t the deformation parameter, and

$$F(z, 0) \equiv f(z) \qquad\qquad (|z| < 1).$$

We require that F map (z, t) continuously into the w-
sphere and reduce to an interior transformation for each
fixed t. Let (α^t) be the characteristic set of F at the
time t. We require that the points of (α^t) vary continuously with t on S, and remain distinct and constant in
number and character, as t varies from 0 to 1, and return respectively when t = 1 to some, but not necessarily
the same, one of the characteristic points of f of like
character. In the case in which (α^t) is independent of t
the deformation of f is termed restricted.
     In this introduction we shall assume that the def-
ormations are restricted. Two interior transformations
f_1 and f_2 which admit a restricted deformation into each
other will be said to be in the same restricted deformation class.
     The invariants J_1. Given an admissible interior
transformation f of S with the characteristic set (α) it
is possible to define n numbers

$$J_i(f, \alpha) \qquad (i = 1, \ldots, n)$$

which are invariant under restricted deformations of f, and which have the property that, if f_1 and f_2 are two such deformations of S with the same characteristic set (α), and if

$$J_i(f_1, \alpha) = J_i(f_2, \alpha) \quad (i = 1, \ldots, n)$$

then f_1 and f_2 are in the same restricted deformation class. If F is a particular interior transformation of S with the characteristic set (α), and f is an arbitrary transformation of this character, then

(34.3) $J_i(f, \alpha) = J_i(F, \alpha) + r_i \quad (i = 1, \ldots, n)$

where r_i is an integer. The r_i's moreover can be pre-scribed as integers, and it can then be shown that there exists an interior transformation of S with the characteristic set (α) and invariants (J) given by (34.3). There is accordingly a countably infinite set of restricted deformation classes with invariants given by (34.3), holding F fast and varying the integers r_i.

In addition it can be shown that there is a meromorphic transformation f in each restricted deformation class, and that two meromorphic transformations f_1 and f_2 in the same restricted deformation classes admit a restricted deformation into each other of meromorphic type, that is, through interior transformations of S which are meromorphic.

Thus one has the remarkable fact that the addition of the interior transformations to the meromorphic transformation does not introduce any new restricted deforma-

tion classes or permit the amalgamation of any classes.
This is not at all obvious, since there are simple domains of definition other than S for which it is not true.
Concerning the topological definition of the invariants J_i a brief comment will be made. One starts with a
simple arc h_i which joins a_0 to a_i and which intersects
no other characteristic point,and considers the image h_i^f
of h_i under f. The invariant J_i is a numerical topological characteristic of h_i^f which is independent of the
choice of the simple arc h_i and of restricted deformations of f. Its definition involves the extensions of
the notions of angular order and order of a locally,
simple, sensed closed curve to a <u>difference</u> <u>order</u> d(k) of
a locally simple sensed arc k, which joins two prescribed
points. The basic notions which we have developed in the
last two sections appear again with a variation appropriate to the case at hand. Like angular orders difference
orders take on a countably infinite set of values.

Covering properties of sequences of meromorphic
transformations of S. One is always concerned with characteristic differences between the theory of meromorphic
and interior transformations. Such differences occur in
the way in which a sequence $[f_k]$ of meromorphic transformations which includes at most one transformation from
each restricted deformation class,covers the w-sphere with
its respective images of S, as compared with a similar
sequence of interior transformations.

We term the sequence $[f_k]$ <u>model</u> if its members are
meromorphic, possess the characteristic set (α), and for
different integers k belong to different restricted deformation classes. We shall be concerned with the set W of
points $w = f_k(z)$, (k = 1, 2, ...), on the w-sphere given
by meromorphic transformations in a model sequence $[f_k]$.
When the characteristic set (α) includes both zeros and
poles the set W covers every point of the w-sphere infinitely often.

No such property holds for a sequence $[f_k]$ of interior transformations in general. In fact, one can
choose an arbitrary closed set H on the w-sphere such
that H does not include w = 0 and w = ∞, and then define
an interior transformation f with the characteristic set
(α) in an arbitrary restricted deformation class, so that
no point w = f(z) for z on S lies on H.

In the case in which the characteristic set contains
zeros and no poles, the set W of a model sequence covers
each point of the w-sphere infinitely many times (w = ∞
excepted) provided f_k does not converge uniformly to zero
on every compact subset of S. More intimate covering
theorems of this character have been established.

These covering properties of model sequences reflect
the "stiffness" of meromorphic functions as compared with
the flexibility of interior transformations. These covering theorems are an immediate consequence of theorems on
"normal families" once the basic analytic properties of
the restricted deformation classes have been discovered.
An extended account of this deformation theory has been
written by Morse and Heins (2) and will presently appear.
The definition and characterization of these deformation
classes leads to a variety of new problems.

References

CARATHÉODORY, C. Conformal representation. Cambridge
University Press, 1932.

FRÉCHET, M. (1) Sur quelques points du calcul fonction-
el. Rendiconti, Circolo matematico di Palermo
22 (1905), pp. 1-74.
(2) Sur une représentation paramétrique intrinsèque
de la courbe continue la plus générale. Journ.
de Math. Sér. 9, 4 (1925), pp. 281-297.

v. KERÉKJÁRTÓ. Vorlesungen über Topologie I. Berlin,
Springer, 1923.

KIANG, TSAI-HAN. Critical points of harmonic functions
and Green's functions in plane regions. Science
Quarterly, National University of Peking 3, pp.
113-123.

KURATOWSKI, C. Théorèmes sur l'homotopie des fonctions
continues de variable complexe et leur rapports
à la Théorie des fonctions analytiques. Fund.
Math. 33 (1945) pp. 316-367.

MONTEL, PAUL. Leçons sur les familles normales de fonctions analytiques. Paris, Gauthier-Villars,
1927.

MORSE, M. (1) The topology of pseudo-harmonic functions.
Duke Math. Jour. 13 (1946), pp. 21-42.
(2) A special parameterization of curves. Bull.,
Amer. Math. Soc. 42 (1936), pp. 915-922.
(3) Singular points of vector fields under general
boundary conditions. Amer. Jour. of Math. 51
(1929), pp. 165-178.

MORSE, M. AND HEINS, M. (1) Topological methods in the
     theory of functions of a single complex variable:
     I. Deformation types of locally simple
         curves. Annals of Math. 46 (1945), pp.
         600-624.
        II. Boundary values and integral character-
             istics of interior transformations and
             pseudo-harmonic functions. Idem, pp.
             625-666.
       III. Causal isomorphisms in the theory of
             pseudo-harmonic functions. Idem, 47
             (1946), pp. 233-274.
     (2) Deformation classes of meromorphic functions
          and their extensions to interior transforma-
          tions. Acta Mathematica, 80 (1947).

OSGOOD, W. F. Lehrbuch der Funktionentheorie I. Berlin,
     Teubner, 1928.

RADÓ, T. Zur Theorie der mehrdeutigen konformen Abbil-
     dungen. Szeged Univ., Acta litt. ac Scient. 1
     (1922), pp. 55-64.

STOILOW. (1) Leçons sur les principes topologiques de
     la théorie des fonctions analytiques. Paris,
     Gauthier-Villars, 1938.
     (2) Du caractère topologique d'un théorème sur les
          fonctions méromorphes. C. R., Acad. des Sci.
          de Paris 190 (1930), pp. 251-253.

TITCHMARSH, E. C. The theory of functions. Oxford, 1932.

WALSH, J. L. The location of the critical points of har-
     monic functions. Proc., Nat. Acad. of Sci. 20
     (1934), pp. 551-554.

WHITNEY, H. (1) Regular families of curves. Annals of
     Math. 34 (1933), pp. 244-270.
     (2) On regular closed curves in the plane. Composi-
          tio Mathematica 4 (1937), pp. 276-284.

WHYBURN, G. T. Analytic topology. Amer. Math. Soc.,
        Coll. Lect. New York, 1942.

GLOSSARY

Angular order p, 63.

Branch point: definition of, 3; order of, 3; partial, 83.

Boundary index I, 38.

Boundary Conditions: A, 12; B, C, 43; generalized, 43; I, 65; II, 81.

Canonical neighborhood: see neighborhood, canonical.

Critical point: definition of, 10, 21, 24; multiplicity of, 10, 21.

Curve: regular, 44; p-curve, 99; μ-length of, 100.

Deformations: of interior transformations, 4; of locally simple curves, 108.

Emergent boundary point, 44; tangentially emergent, 54.

Entrant covering, 54.

Entrant boundary point, 44; tangentially entrant, 54.

Fréchet distance, 49.

Milton Keynes UK
Ingram Content Group UK Ltd.
UKHW022337260224
438516UK00001B/40